Albert Einstein

THE THEORY OF Relativity & other essays

BARNES & NOBLE BOOKS
NEW YORK

Published by MJF Books
Fine Communications
322 Eighth Avenue
New York, NY 10001

The Theory of Relativity and Other Essays
Library of Congress Catalog Card Number 97-75622
ISBN 1-56731-247-0

This edition published by arrangement with Carol Publishing Group.

Manufactured in the United States of America on acid-free paper

MJF Books and the MJF colophon are trademarks of Fine Creative Media, Inc.

QM 10 9 8 7 6 5

CONTENTS

THE THEORY OF RELATIVITY
(and Other Essays)

1. *THE THEORY OF RELATIVITY*

MATHEMATICS DEALS exclusively with the relations of concepts to each other without consideration of their relation to experience. Physics too deals with mathematical concepts; however, these concepts attain physical content only by the clear determination of their relation to the objects of experience. This in particular is the case for the concepts of motion, space, time.

The theory of relativity is that physical theory which is based on a consistent physical interpretation of these three concepts. The name "theory of relativity" is connected with the fact that motion from the point of view of possible experience always appears as the *relative* motion of one object with respect to another (e.g., of a car with respect to the ground, or the earth with respect to the sun and the fixed stars). Motion is never observable as "motion with respect to space" or, as it has been expressed, as "absolute motion." The "principle of relativity" in its widest sense is contained in the statement: The totality of physical phenomena is of such a character that it gives no basis for the introduction of the concept of "absolute motion"; or shorter but less precise: There is no absolute motion.

It might seem that our insight would gain little from such a negative statement. In reality, however, it is a strong restriction for the (conceivable) laws of nature. In this sense there exists an analogy between the theory of relativity and thermodynamics. The latter too is based on a negative statement: "There exists no perpetuum mobile."

The development of the theory of relativity proceeded in two steps, "special theory of relativity" and "general theory of relativity." The latter presumes the validity of the former as a limiting case and is its consistent continuation.

A. *Special theory of relativity.*

Physical interpretation of space and time in classical mechanics.

Geometry, from a physical standpoint, is the totality of laws according to which rigid bodies mutually at rest can be placed with respect to each other (e.g., a triangle consists of three rods whose ends touch permanently). It is assumed that with such an interpretation the Euclidean laws are valid. "Space" in this interpretation is in principle an infinite rigid body (or skeleton) to which the position of all other bodies is related (body of reference). Analytic geometry (Descartes) uses as the body of reference, which represents space, three mutually perpendicular rigid rods on which the "coordinates" (x, y, z) of space points are measured in the known manner as perpendicular projections (with the aid of a rigid unit-measure).

Physics deals with "events" in space and time. To each event belongs, besides its place coordinates x, y, z, a time value t. The latter was considered measurable by a clock (ideal periodic process) of negligible spatial extent. This clock C is to be considered at rest at one point of the coordinate system, e.g., at the coordinate origin ($x = y = z = O$). The time of an event taking place at a point P (x, y, z) is then defined as the time shown on the clock C simultaneously with the event. Here the concept "simultaneous" was assumed as physically meaningful without special definition. This is a lack of exactness which seems harmless only since with the help of light (whose velocity is practically infinite from the point of view of daily experience) the simultaneity of spatially distant events can apparently be decided immediately.

The special theory of relativity removes this lack of precision by defining simultaneity physically with the use of light signals. The time t of the event in P is the reading of the clock C at the time of arrival of a light signal emitted from the event, corrected with respect to the time needed for the light signal to travel the distance. This correction presumes (postulates) that the velocity of light is constant.

This definition reduces the concept of simultaneity of spatially distant events to that of the simultaneity of events happening at the same place (coincidence), namely the arrival of the light signal at C and the reading of C.

Classical mechanics is based on Galileo's principle: A body is in rectilinear and uniform motion as long as other bodies do not act on it. This statement cannot be valid for arbitrary moving systems of coordinates. It can claim validity only for so-called "inertial systems." Inertial systems are in rectilinear and uniform motion with respect to each other. In classical physics laws claim validity only with respect to all inertial systems (special principle of relativity).

It is now easy to understand the dilemma which has led to the special theory of relativity. Experience and theory have gradually led to the conviction that light in empty space always travels with the same velocity c independent of its color and the state of motion of the source of light (principle of the constancy of the velocity of light—in the following referred to as "L-principle"). Now elementary intuitive considerations seem to show that the same light ray *cannot* move with respect to all inertial systems with the same velocity c. The L-principle seems to contradict the special principle of relativity.

It turns out, however, that this contradiction is only an apparent one which is based essentially on the prejudice about the absolute character of time or rather of the simultaneity of distant events. We just saw that x, y, z and t of an event can, for the moment, be defined only with respect to a certain chosen system of coordinates (inertial system). The transformation of the x, y, z, t of events which has to be carried out with the passage from one inertial system to another (coordinate transformation), is a problem which cannot be solved without special physical assumptions. However, the following postulate is exactly sufficient for a solution: *The L-principle holds for all inertial systems* (application of the special principle of relativity to the L-principle). The transformations thus defined, which are linear in x, y, z, t, are called Lorentz transformations. Lorentz transformations are formally characterized by the demand that the expression

$$dx^2 + dy^2 + dz^2 - c^2dt^2,$$

which is formed from the coordinate differences dx, dy, dz, dt of two infinitely close events, be invariant (i.e., that through the transformation it goes over into the *same* expression formed from the coordinate differences in the new system).

With the help of the Lorentz transformations the special principle of relativity can be expressed thus: The laws of nature are invariant with respect to Lorentz-transformations (i.e., a law of nature does not change its form if one introduces into it a new inertial system with the help of a Lorentz-transformation on x, y, z, t).

The special theory of relativity has led to a clear understanding of the physical concepts of space and time and in connection with this to a recognition of the behavior of moving measuring rods and clocks. It has in principle removed the concept of absolute simultaneity and thereby also that of instantaneous action at a distance in the sense of Newton. It has shown how the law of motion must be modified in dealing with motions that are not negligibly small as compared with the velocity of light. It has led to a formal clarification of Maxwell's equations of the electromagnetic field; in particular it has led to an understanding of the essential oneness of the electric and the magnetic field. It has unified the laws of conservation of momentum and of energy into one single law and has demonstrated the equivalence of mass and energy. From a formal point of view one may characterize the achievement of the special theory of relativity thus: it has shown generally the role which the universal constant c (velocity of light) plays in the laws of nature and has demonstrated that there exists a close connection between the form in which time on the one hand and the spatial coordinates on the other hand enter into the laws of nature.

B. *General theory of relativity.*

The special theory of relativity retained the basis of classical mechanics in one fundamental point, namely the statement: The laws of nature are valid only with respect to inertial systems. The "permissible" transformations for the coordinates (i.e., those which leave the form of the laws unchanged) are *exclusively* the (linear) Lorentz-transformations. Is this restriction really founded in physical facts? The following argument convincingly denies it.

Principle of equivalence. A body has an inertial mass (resistance to acceleration) and a heavy mass (which determines the weight of the body in a given gravitational field, e.g., that at the surface of the earth). These two quantities, so different according to their definition, are according to experience measured by one and the same number. There must be a deeper reason for this. The fact can also be described thus: In a gravitational field different masses receive the same acceleration. Finally, it can also be expressed thus: Bodies in a gravitational field behave as in the absence of a gravitational field if, in the latter case, the system of reference used is a uniformly accelerated coordinate system (instead of an inertial system).

There seems, therefore, to be no reason to ban the following interpretation of the latter case. One considers the system as being "at rest" and considers the "apparent" gravitational field which exists with respect to it as a "real" one. This gravitational field "generated" by the acceleration of the coordinate system would of course be of unlimited extent in such a way that it could not be caused by gravitational masses in a finite region; however, if we are looking for a field-like theory, this fact need not deter us. With this interpretation the inertial system loses its meaning and one has an "explanation" for the equality of heavy and inertial mass (the same property of matter appears as weight or as inertia depending on the mode of description).

Considered formally, the admission of a coordinate system which is accelerated with respect to the original "inertial" coordinates means the admission of non-linear coordinate transformations, hence a mighty enlargement of the idea of invariance, i.e., the principle of relativity.

First, a penetrating discussion, using the results of the special theory of relativity, shows that with such a generalization the coordinates can no longer be interpreted directly as the results of measurements. Only the coordinate difference together with the field quantities which describe the gravitational field determine measurable distances between events. After one has found oneself forced to admit non-linear coordinate transformations as transformations between equivalent coordinate systems, the simplest demand appears to admit all continuous coordinate transformations (which form a group), i.e., to admit arbitrary curvilinear coordinate systems in which the fields are described by regular functions (general principle of relativity).

Now it is not difficult to understand why the general principle of relativity (*on the basis of the equivalence principle*) has led to a theory of gravitation. There is a special kind of space whose physical structure (field) we can presume as precisely known on the basis of the special theory of relativity. This is empty space without electromagnetic field and without matter. It is completely determined by its "metric" property: Let dx_0, dy_0, dz_0, dt_0 be the coordinate differences of two infinitesimally near points (events); then

$$ds^2 = dx_0^2 + dy_0^2 + dz_0^2 - c^2dt_0^2 \qquad (1)$$

is a measurable quality which is independent of the special choice of the inertial system. If one introduces in this space the new coordinates x_1, x_2, x_3, x_4 through a general transformation of coordinates, then the quantity

ds^2 for the same pair of points has an expression of the form

(2) $ds^2 = \Sigma g_{ik} dx^i dx^k$ (summed for i and k from 1 to 4)

where $g_{ik} = g_{ki}$. The g_{ik} which form a "symmetric tensor" and are continuous functions of $x_1, \ldots x_4$ then describe according to the "principle of equivalence" a gravitational field of a special kind (namely one which can be retransformed to the form [1]). From Riemann's investigations on metric spaces the mathematical properties of this g_{ik} field can be given exactly ("Riemann-condition"). However, what we are looking for are the equations satisfied by "general" gravitational fields. It is natural to assume that they too can be described as tensor-fields of the type g_{ik}, which in general do *not* admit a transformation to the form (1), i.e., which do not satisfy the "Riemann condition," but weaker conditions, which, just as the Riemann condition, are independent of the choice of coordinates (i.e., are generally invariant). A simple formal consideration leads to weaker conditions which are closely connected with the Riemann condition. These conditions are the very equations of the pure gravitational field (on the outside of matter and at the absence of an electromagnetic field).

These equations yield Newton's equations of gravitational mechanics as an approximate law and in addition certain small effects which have been confirmed by observation (deflection of light by the gravitational field of a star, influence of the gravitational potential on the frequency of emitted light, slow rotation of the elliptic circuits of planets—perihelion motion of the planet Mercury). They further yield an explanation for the expanding motion of galactic systems, which is manifested by the red-shift of the light omitted from these systems.

The general theory of relativity is as yet incomplete insofar as it has been able to apply the general principle of relativity satisfactorily only to gravitational fields, but not to the total field. We do not yet know with certainty, by what mathematical mechanism the total field in space is to be described and what the general invariant laws are to which this total field is subject. One thing, however, seems certain: namely, that the general principle of relativity will prove a necessary and effective tool for the solution of the problem of the total field.

2. $E = MC^2$

In order to understand the law of the equivalence of mass and energy, we must go back to two conservation or "balance" principles which, independent of each other, held a high place in pre-relativity physics. These were the principle of the conservation of energy and the principle of the conservation of mass. The first of these, advanced by Leibnitz as long ago as the seventeenth century, was developed in the nineteenth century essentially as a corollary of a principle of mechanics.

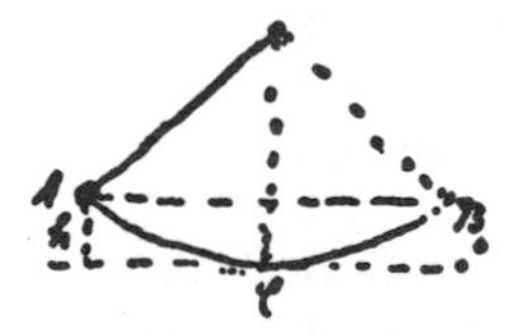

Drawing from Dr. Einstein's manuscript.

Consider, for example, a pendulum whose mass swings back and forth between the points A and B. At these points the mass m is higher by the amount h than it is at C, the lowest point of the path (see drawing). At C, on the other hand, the lifting height has disappeared and instead of it the mass has a velocity v. It is as though the lifting height could be converted entirely into velocity, and vice versa. The exact relation would be expressed as $mgh = \frac{m}{2} v^2$, with g representing the acceleration of gravity. What is interesting here is that this relation is independent of both the length of the pendulum and the form of the path through which the mass moves.

The significance is that something remains constant throughout the process, and that something is energy. At A and at B it is an energy of position, or "potential" energy; at C it is an energy of motion, or "kinetic" energy. If this concept is correct, then the sum $mgh + m\frac{v^2}{2}$ must have the same value for any position of the pendulum, if h is understood to represent the height above C, and v the velocity at that point in the pendulum's path. And such is found to be actually the case. The generalization of this principle gives us the law of the conservation of mechanical energy. But what happens when friction stops the pendulum?

The answer to that was found in the study of heat phenomena. This study, based on the assumption that heat is an indestructible substance which flows from a warmer to a colder object, seemed to give us a principle of the "conservation of heat." On the other hand, from time immemorial it has been known that heat could be produced by friction, as in the fire-making drills of the Indians. The physicists were for long unable to account for this kind of heat "production." Their difficulties were overcome only when it was successfully established that, for any given amount of heat produced by friction, an exactly proportional amount of energy had to be expended. Thus did we arrive at a principle of the "equivalence of work and heat." With our pendulum, for example, mechanical energy is gradually converted by friction into heat.

In such fashion the principles of the conservation of mechanical and thermal energies were merged into one. The physicists were thereupon persuaded that the conservation principle could be further extended to take in chemical and electromagnetic processes—in short, could be applied to all fields. It appeared that in our physical system there was a sum total of energies that remained constant through all changes that might occur.

Now for the principle of the conservation of mass. Mass is defined by the resistance that a body opposes to its acceleration (inert mass). It is also measured by the weight of the body (heavy mass). That these two radically different definitions lead to the same value for the mass of a body is, in itself, an astonishing fact. According to the principle—namely, that masses remain unchanged under any physical or chemical changes—the mass appeared to be the essential (because unvarying) quality of matter. Heating, melting, vaporization, or combining into chemical compounds would not change the total mass.

Physicists accepted this principle up to a few decades ago. But it proved inadequate in the face of the special theory of relativity. It was therefore merged with the energy principle—just as, about 60 years before, the principle of the conservation of mechanical energy had been combined with the principle of the conservation of heat. We might say that the principle of the conservation of energy, having previously swallowed up that of the conservation of heat, now proceeded to swallow that of the conservation of mass—and holds the field alone.

It is customary to express the equivalence of mass and energy (though somewhat inexactly) by the formula $E = mc^2$, in which c represents the velocity of light, about 186,000 miles per second. E is the energy that is contained in a stationary body; m is its mass. The energy that belongs to the mass m is equal to this mass, multiplied by the square of the enormous speed of light—which is to say, a vast amount of energy for every unit of mass.

But if every gram of material contains this tremendous energy, why did it go so long unnoticed? The answer is simple enough: so long as none of the energy is given off externally, it cannot be observed. It is as though a man who is fabulously rich should never spend or give away a cent; no one could tell how rich he was.

Now we can reverse the relation and say that an increase of E in the amount of energy must be accompanied by an increase of $\frac{E}{c^2}$ in the mass. I can easily supply energy to the mass—for instance, if I heat it by 10 degrees. So why not measure the mass increase, or weight increase, connected with this change? The trouble here is that in the mass increase the enormous factor c^2 occurs in the denominator of the fraction. In such a case the increase is too small to be measured directly; even with the most sensitive balance.

For a mass increase to be measurable, the change of energy per mass unit must be enormously large. We know of only one sphere in which such amounts of energy per mass unit are released: namely, radioactive disintegration. Schematically, the process goes like this: An atom of the mass M splits into two atoms of the mass M′ and M″, which separate with tremendous kinetic energy. If we imagine these two masses as brought to rest—that is, if we take this energy of motion from them—then, considered together, they are essentially poorer in energy than was the original atom. According to the equivalence principle, the mass sum $M' + M''$ of the disintegration products must also be somewhat smaller than the original mass M of the disintegrating atom—in contradiction to the old principle of the conservation of mass. The relative difference of the two is on the order of $1/10$ of one percent.

Now, we cannot actually weigh the atoms individually. However, there are indirect methods for measuring their weights exactly. We can likewise determine the kinetic energies that are transferred to the disintegration products M′ and M″. Thus it has become possible to test and confirm the equivalence formula. Also, the law permits us to calculate in advance, from precisely determined atom weights, just how much energy will be released with any atom disintegration we have in mind. The law says nothing, of course, as to whether—or how—the disintegration reaction can be brought about.

What takes place can be illustrated with the help of our rich man. The atom M is a rich miser who, during his life, gives away no money (*energy*). But in his will he bequeaths his fortune to his sons M′ and M″, on condition that they give to the community a small amount, less than one thousandth of the whole estate (*energy or mass*). The sons together have somewhat less than the father had (*the mass sum $M' + M''$ is somewhat smaller than the mass M of the radioactive atom*). But the part given to the community, though relatively small, is still so enormously large (*considered as kinetic energy*) that it brings with it a great threat of evil. Averting that threat has become the most urgent problem of our time.

3. *PHYSICS AND REALITY*

§ 1. General Consideration Concerning the Method of Science

It has often been said, and certainly not without justification, that the man of science is a poor philosopher. Why then should it not be the right thing for the physicist to let the philosopher do the philosophizing? Such might indeed be the right thing at a time when the physicist believes he has at his disposal a rigid system of fundamental concepts and fundamental laws which are so well established that waves of doubt can not reach them; but, it can not be right at a time when the very foundations of physics itself have become problematic as they are now. At a time like the present, when experience forces us to seek a newer and more solid foundation, the physicist cannot simply surrender to the philosopher the critical contemplation of the theoretical foundations; for, he himself knows best, and feels more surely where the shoe pinches. In looking for a new foundation, he must try to make clear in his own mind just how far the concepts which he uses are justified, and are necessities.

The whole of science is nothing more than a refinement of everyday thinking. It is for this reason that the critical thinking of the physicist cannot possibly be restricted to the examination of the concepts of his own specific field. He cannot proceed without considering critically a much more difficult problem, the problem of analyzing the nature of everyday thinking.

On the stage of our subconscious mind appear in colorful succession sense experiences, memory pictures of them, representations and feelings. In contrast to psychology, physics treats directly only of sense experiences and of the "understanding" of their connection. But even the concept of the "real external world" of everyday thinking rests exclusively on sense impressions.

Now we must first remark that the differentiation between sense impressions and representations is not possible; or, at least it is not possible with absolute certainty. With the discussion of this problem, which affects also the notion of reality, we will not concern ourselves but we shall take the existence of sense experiences as given, that is to say as psychic experiences of special kind.

I believe that the first step in the setting of a "real external world" is the formation of the concept of bodily objects and of bodily objects of various kinds. Out of the multitude of our sense experiences we take, mentally and arbitrarily, certain repeatedly occurring complexes of sense impression (partly in conjunction with sense impressions which are interpreted as signs for sense experiences of others), and we attribute to them a meaning —the meaning of the bodily object. Considered logically this concept is not identical with the totality of sense impressions referred to; but it is an arbitrary creation of the human (or animal) mind. On the other hand, the concept owes its meaning and its justification exclusively to the totality of the sense impressions which we associate with it.

The second step is to be found in the fact that, in our thinking (which determines our expectation), we attribute to this concept of the bodily object a significance, which is to a high degree independent of the sense impression which originally gives rise to it. This is what we mean when we attribute to the bodily object "a real existence." The justification of such a setting rests exclusively on the fact that, by means of such concepts and mental relations between them, we are able to orient ourselves in the labyrinth of sense impressions. These notions and relations, although free statements of our thoughts, appear to us as stronger and more unalterable than the individual sense experience itself, the character of which as anything other than the result of an illusion or hallucination is never completely guaranteed. On the other hand, these concepts and relations, and indeed the setting of real objects and, generally speaking, the existence of "the real world," have justification only in so far as they are connected with sense impressions between which they form a mental connection.

The very fact that the totality of our sense experiences is such that by means of thinking (operations with concepts, and the creation and use of definite functional relations between them, and the coordination of sense experiences to these concepts) it can be put in order, this fact is one which leaves us in awe, but which we shall never understand. One may say "the eternal mystery of the world is its comprehensibility." It is one of the great realizations of Immanuel Kant that the setting up of a real external world would be senseless without this comprehensibility.

In speaking here concerning "comprehensibility," the expression is used in its most modest sense. It implies: the production of some sort of order among sense impressions, this order being produced by the creation of general concepts, relations between these concepts, and by relations between the concepts and sense experience, these relations being determined in any possible manner.

It is in this sense that the world of our sense experiences is comprehensible. The fact that it is comprehensible is a miracle.

In my opinion, nothing can be said concerning the manner in which the concepts are to be made and connected, and how we are to coordinate them to the experiences. In guiding us in the creation of such an order of sense experiences, success in the result is alone the determining factor. All that is necessary is *the statement* of a set of rules, since without such rules the acquisition of knowledge in the desired sense would be impossible. One may compare these rules with the rules of a game in which, while the rules themselves are arbitrary, it is their rigidity alone which makes the game possible. However, the fixation will never be final. It will have validity only for a special field of application (i.e. there are no final categories in the sense of Kant).

The connection of the elementary concepts of everyday thinking with complexes of sense experiences can only be comprehended intuitively and it is unadaptable to scientifically logical fixation. The totality of these connections—none of which is expressible in notional terms—is the only thing which differentiates the great building which is science from a logical but empty scheme of concepts. By means of these connections, the purely notional theorems of science become statements about complexes of sense experiences.

We shall call "primary concepts" such concepts as are directly and intuitively connected with typical complexes of sense experiences. All other notions are—from the physical point of view—possessed of meaning, only in so far as they are connected, by theorems, with the primary notions. These theorems are partially definitions of the concepts (and of the statements derived logically from them) and partially theorems not derivable from the definitions, which express at least indirect relations between the "primary concepts," and in this way between sense experiences. Theorems of the latter kind are "statements about reality" or laws of nature, i.e. theorems which have to show their usefulness when applied to sense experiences comprehended by primary concepts. The question as to which of the theorems shall be considered as definitions and which as natural laws will depend largely upon the chosen representation. It really becomes absolutely necessary to make this differentiation only when one examines the degree to which the whole system of concepts considered is not empty from the physical point of view.

Stratification of the Scientific System

The aim of science is, on the one hand, a comprehension, as *complete* as possible, of the connection between the sense experiences in their totality, and, on the other hand, the accomplishment of this aim *by the use of a minimum of primary concepts and relations.* (Seeking, as far as possible, logical unity in the world picture, i.e. paucity in logical elements.)

Science concerns the totality of the primary concepts, i.e. concepts directly connected with sense experiences, and theorems connecting them. In its first stage of development, science does not contain anything else. Our everyday thinking is satisfied on the whole with this level. Such a state of affairs cannot, however, satisfy a spirit which is really scientifically minded; because, the totality of concepts and relations obtained in this manner is utterly lacking in logical unity. In order to supplement this deficiency, one invents a system poorer in concepts and relations, a system retaining the primary concepts and relations of the "first layer" as logically derived concepts and relations. This new "secondary system" pays for its higher logical unity by having, as its own elementary concepts (concepts of the second layer), only those which are no longer directly connected with complexes of sense experiences. Further striving for logical unity brings us to a tertiary system, still poorer in concepts and relations, for the deduction of the concepts and relations of the secondary (and so indirectly of the primary) layer. Thus the story goes on until we have arrived at a system of the greatest conceivable unity, and of the greatest poverty of concepts of the logical foundations, which are still compatible with the observation made by our senses. We do not know whether or not this ambition will ever result in a definite system. If one is asked for his opinion, he is inclined to answer no. While wrestling with the problems, however, one will never give up the hope that this greatest of all aims can really be attained to a very high degree.

An adherent to the theory of abstraction or induction might call our layers "degrees of abstraction"; but, I do not consider it justifiable to veil the logical independence of the concept from the sense experiences. The relation is not analogous to that of soup to beef but rather of wardrobe number to overcoat.

The layers are furthermore not clearly separated. It is not even absolutely clear which concepts belong to the primary layer. As a matter of fact, we are dealing with freely formed concepts, which, with a certainty sufficient for practical use, are intuitively connected with complexes of sense experiences in such a manner that, in any given case of experience, there is no uncertainty as to the applicability or non-applicability of the statement. The essential thing is the aim to represent the multitude of concepts and theorems, close to experience, as theorems, logically deduced and belonging to a basis, as narrow as possible, of fundamental concepts and fundamental relations which themselves can be chosen freely (axioms). The liberty of choice, however, is of a special kind; it is not in any way similar to the liberty of a writer of fiction. Rather, it is similar to that of a man engaged in solving a well designed word puzzle. He may, it is true, propose any word as the solution; but, there is only *one* word which really solves the puzzle in all its forms. It is an outcome of faith that nature—as she is perceptible to our five senses—takes the character of such a well formulated puzzle. The successes reaped up to now by science do, it is true, give a certain encouragement for this faith.

The multitude of layers discussed above corresponds to the several stages of progress which have resulted from the struggle for unity in the course of development.

As regards the final aim, intermediary layers are only of temporary nature. They must eventually disappear as irrelevant. We have to deal, however, with the science of today, in which these strata represent problematic partial successes which support one another but which also threaten one another, because today's systems of concepts contain deep seated incongruities which we shall meet later on.

It will be the aim of the following lines to demonstrate what paths the constructive human mind has entered, in order to arrive at a basis of physics which is logically as uniform as possible.

§ 2. Mechanics and the Attempts to Base All Physics upon It

An important property of our sense experiences, and, more generally, of all of our experience, is its time-like order. This kind of order leads to the mental conception of a subjective time, an ordinating scheme for our experience. The subjective time leads then through the concept of the bodily object and of space, to the concept of objective time, as we shall see later on.

Ahead of the notion of objective time there is, however, the concept of space; and, ahead of the latter we find the concept of the bodily object. The latter is directly connected with complexes of sense experiences. It has been pointed out that one property which is characteristic of the notion "bodily object" is the property which provides that we coordinate to it an existence, independent of (subjective) time, and independent of the fact that it is perceived by our senses. We do this in spite of the fact that we perceive temporal alterations in it. Poincaré has justly emphasized the fact that we distinguish two kinds of alterations of the bodily object, "changes of state" and "changes of position." The latter, he remarked, are alterations which we can reverse by arbitrary motions of our bodies.

That there are bodily objects to which we have to ascribe, within a certain sphere of perception, no alteration of state, but only alterations of position, is a fact of fundamental importance for the formation of the concept of space (in a certain degree even for the justification of the notion of the bodily object itself). Let us call such an object "practically rigid."

If, as the object of our perception, we consider simultaneously (i.e. as a single unit) two practically rigid bodies, then there exist for this ensemble such alterations as can *not* possibly be considered as changes of position of the whole, notwithstanding the fact that this is the case for each one of the two constituents. This leads to the notion of "change of relative position" of the two objects; and, in this way, also to the notion of "relative position" of the two objects. It is found moreover that among the relative positions, there is one of a specific kind which we designate as "Contact." [1] Permanent contact of two bodies in three or more "points" means that they are united as a quasi rigid compound body. It is permissible to say that the second body forms then a (quasi rigid) continuation on the first body and may, in its turn, be continued quasi rigidly. The possibility of the quasi rigid continuation of a body is unlimited. The real essence of the conceivable quasi rigid continuation of a body B_0 is the infinite "space" determined by it.

In my opinion, the fact that every bodily object situated in any arbitrary manner can be put into contact with the quasi rigid continuation of a predetermined and chosen body B_0 (body of relation), this fact is the empirical basis of our conception of space. In pre-scientific thinking, the solid earth's crust plays the role of B_0 and its continuation. The very name geometry indicates that the concept of space is psychologically connected with the earth as an assigned body.

The bold notion of "space" which preceded all scientific geometry transformed our mental concept of the relations of positions of bodily objects into the notion of

[1] It is in the nature of things that we are able to talk about these objects only by means of concepts of our own creation, concepts which themselves are not subject to definition. It is essential, however, that we make use only of such concepts concerning whose coordination to our experience we feel no doubt.

the position of these bodily objects in "space." This, of itself, represents a great formal simplification. Through this concept of space one reaches, moreover, an attitude in which any description of position is admittedly a description of contact; the statement that a point of a bodily object is located at a point P of space means that the object touches the point P of the standard body of reference B_0 (supposed appropriately continued) at the point considered.

In the geometry of the Greeks, space plays only a qualitative role, since the position of bodies in relation to space is considered as given, it is true, but is not described by means of numbers. Descartes was the first to introduce this method. In his language, the whole content of Euclidian geometry can axiomatically be founded upon the following statements: (1) Two specified points of a rigid body determine a distance. (2) We may coordinate triplets of numbers X_1, X_2, X_3, to points of space in such a manner that for every distance $P' - P''$ under consideration, the coordinates of whose end points are X_1', X_2', X_3'; X_1'', X_2'', X_3'', the expression

$$S^2 = (X_1'' - X_1')^2 + (X_2'' - X_2')^2 + (X_3'' - X_3')^2$$

is independent of the position of the body, and of the positions of any and all other bodies.

The (positive) number S means the length of the stretch, or the distance between the two points P' and P'' of space (which are coincident with the points P' and P'' of the stretch).

The formulation is chosen, intentionally, in such a way that it expresses clearly, not only the logical and axiomatic, but also the empirical content of Euclidian geometry. The purely logical (axiomatic) representation of Euclidian geometry has, it is true, the advantage of greater simplicity and clarity. It pays for this, however, by renouncing representation of the connection between the notional construction and the sense experience upon which connection, alone, the significance of geometry for physics rests. The fatal error that the necessity of thinking, preceding all experience, was at the basis of

Euclidian geometry and the concept of space belonging to it, this fatal error arose from the fact that the empirical basis, on which the axiomatic construction of Euclidian geometry rests, had fallen into oblivion.

In so far as one can speak of the èxistence of rigid bodies in nature, Euclidian geometry is a physical science, the usefulness of which must be shown by application to sense experiences. It relates to the totality of laws which must hold for the relative positions of rigid bodies independently of time. As one may see, the physical notion of space also, as originally used in physics, is tied to the existence of rigid bodies.

From the physicist's point of view, the central importance of Euclidian geometry rests in the fact that its laws are independent of the specific nature of the bodies whose relative positions it discusses. Its formal simplicity is characterized by the properties of homogeneity and isotropy (and the existence of similar entities).

The concept of space is, it is true, useful, but not indispensable for geometry proper, i.e. for the formulation of rules about the relative positions of rigid bodies. In opposition to this, the concept of objective time, without which the formulation of the fundamentals of classical mechanics is impossible, is linked with the concept of the spacial continuum.

The introduction of objective time involves two statements which are independent of each other.

(1) The introduction of the objective local time by connecting the temporal sequence of experiences with the indications of a "clock," i.e. of a closed system with periodical occurrence.

(2) The introduction of the notion of objective time for the happenings in the whole space, by which notion alone the idea of local time is enlarged to the idea of time in physics.

Note concerning (1). As I see it, it does not mean a "petitio principii" if one puts the concept of periodical occurrence ahead of the concept of time, while one is concerned with the clarification of the origin and of the empirical content of the concept of time. Such a conception corresponds exactly to the precedence of the concept of the rigid (or quasi rigid) body in the interpretation of the concept of space.

Further discussion of (2). The illusion which prevailed prior to the enunciation of the theory of relativity —that, from the point of view of experience the meaning of simultaneity in relation to happenings distant in space and consequently that the meaning of time in physics is a priori clear—this illusion had its origin in the fact that in our everyday experience, we can neglect the time of propagation of light. We are accustomed on this account to fail to differentiate between "simultaneously seen" and "simultaneously happening"; and, as a result the difference between time and local time fades away.

The lack of definiteness which, from the point of view of empirical importance, adheres to the notion of time in classical mechanics was veiled by the axiomatic representation of space and time as things given independently of our senses. Such a use of notions—independent of the empirical basis, to which they owe their existence —does not necessarily damage science. One may however easily be led into the error of believing that these notions, whose origin is forgotten, are necessary and unalterable accompaniments to our thinking, and this error may constitute a serious danger to the progress of science.

It was fortunate for the development of mechanics and hence also for the development of physics in general, that the lack of definiteness in the concept of objective time remained obscured from the earlier philosophers as regards its empirical interpretation. Full of confidence in the real meaning of the space-time construction they developed the foundations of mechanics which we shall characterize, schematically, as follows:

(*a*) Concept of a material point: a bodily object which —as regards its position and motion—can be described with sufficient exactness as a point with coordinates X_1, X_2, X_3. Description of its motion (in relation to the "space" B_0) by giving X_1, X_2, X_3, as functions of the time.

(*b*) Law of inertia: the disappearance of the components of acceleration for the material point which is sufficiently far away from all other points.

(*c*) Law of motion (for the material point): Force = mass × acceleration.

(*d*) Laws of force (actions and reactions between material points).

In this (*b*) is nothing more than an important special case of (*c*). A real theory exists only when the laws of force are given. The forces must in the first place only obey the law of equality of action and reaction in order that a system of points—permanently connected to each other—may behave like *one* material point.

These fundamental laws, together with Newton's law for gravitational force, form the basis of the mechanics of celestial bodies. In this mechanics of Newton, and in contrast to the above conceptions of space derived from rigid bodies, the space B_0 enters in a form which contains a new idea; it is not for every B_0 that validity is required (for a given law of force) by (*b*) and (*c*), but only for a B_0 in the appropriate condition of motion (inertial system). On account of this fact, the coördinate space acquired an independent physical property which is not contained in the purely geometrical notion of space, a circumstance which gave Newton considerable food for thought (pail-experiment).[2]

Classical mechanics is only a general scheme; it becomes a theory only by explicit indication of the force laws (*d*) as was done so very successfully by Newton for celestial mechanics. From the point of view of the aim of the greatest logical simplicity of the foundations, this theoretical method is deficient in so far as the laws of force cannot be obtained by logical and formal considera-

[2] This defect of the theory could only be eliminated by such a formulation of mechanics as would command validity for all B_0. This is one of the steps which lead to the general theory of relativity. A second defect, also eliminated only by the introduction of the general theory of relativity, lies in the fact that there is no reason given by mechanics itself for the equality of the gravitational and inertial mass of the material point.

tions, so that their choice is *a priori* to a large extent arbitrary. Also Newton's gravitation law of force is distinguished from other conceivable laws of force exclusively by its *success*.

In spite of the fact that, today, we know positively that classical mechanics fails as a foundation dominating all physics, it still occupies the center of all of our thinking in physics. The reason for this lies in the fact that, regardless of important progress reached since the time of Newton, we have not yet arrived at a new foundation of physics concerning which we may be certain that the whole complexity of investigated phenomena, and of partial theoretical systems of a successful kind, could be deduced logically from it. In the following lines I shall try to describe briefly how the matter stands.

First we try to get clearly in our minds how far the system of classical mechanics has shown itself adequate to serve as a basis for the whole of physics. Since we are dealing here only with the foundations of physics and with its development, we need not concern ourselves with the purely *formal* progresses of mechanics (equation of Lagrange, canonical equations, etc.). *One* remark, however, appears indispensable. The notion "material point" is fundamental for mechanics. If now we seek the mechanics of a bodily object which itself can *not* be treated as a material point—and strictly speaking every object "perceptible to our senses" is of this category—then the question arises: How shall we imagine the object to be built up out of material points, and what forces must we assume as acting between them? The formulation of this question is indispensable, if mechanics is to pretend to describe the object *completely*.

It is natural to the tendency of mechanics to assume these material points, and the laws of forces acting between them, as invariable, since time alterations would lie outside of the scope of mechanical explanation. From this we can see that classical mechanics must lead us to an atomistic construction of matter. We now realize, with special clarity, how much in error are those theorists who believe that theory comes inductively from experience.

Even the great Newton could not free himself from this error ("Hypotheses non fingo").[8]

In order to save itself from becoming hopelessly lost in this line of thought (atomistic), science proceeded first in the following manner. The mechanics of a system is determined if its potential energy is given as a function of its configuration. Now, if the acting forces are of such a kind as to guarantee maintenance of certain qualities of order of the system's configuration, then the configuration may be described with sufficient accuracy by a relatively small number of configuration variables q_r; the potential energy is considered only insofar as it is dependent upon *these* variables (for instance, description of the configuration of a practically rigid body by six variables).

A second method of application of mechanics, which avoids the consideration of a subdivision of matter down to "real" material points, is the mechanics of so-called continuous media. This mechanics is characterized by the fiction that the density of matter and speed of matter is dependent in a continuous manner upon coordinates and time, and that the part of the interactions not explicitly given can be considered as surface forces (pressure forces) which again are continuous functions of location. Herein we find the hydrodynamic theory, and the theory of elasticity of solid bodies. These theories avoid the explicit introduction of material points by fictions which, in the light of the foundation of classical mechanics, can only have an approximate significance.

In addition to their great *practical* significance, these categories of science have—by enlargement of the mathematical world of ideas—created those formal auxiliary instruments (partial differential equations) which have been necessary for the subsequent attempts at formulating the total scheme of physics in a manner which is new as compared with that of Newton.

These two modes of application of mechanics belong to the so-called "phenomenological" physics. It is characteristic of this kind of physics that it makes as much

[8] "I make no hypotheses."

use as possible of concepts which are close to experience but which, for this reason, have to give up, to a large degree, unity in the foundations. Heat, electricity and light are described by special variables of state and constants of matter other than the mechanical state; and to determine all of these variables in their relative dependence was a rather empirical task. Many contemporaries of Maxwell saw in such a manner of presentation the ultimate aim of physics, which they thought could be obtained purely inductively from experience on account of the relative closeness of the concepts used to the experience. From the point of view of theories of knowledge St. Mill and E. Mach took their stand approximately on this ground.

According to my belief, the greatest achievement of Newton's mechanics lies in the fact that its consistent application has led beyond this phenomenological representation, particularly in the field of heat phenomena. This occurred in the kinetic theory of gases and, in a general way, in statistical mechanics. The former connected the equation of state of the ideal gases, viscosity, diffusion and heat conductivity of gases and radiometric phenomena of gases, and gave the logical connection of phenomena which, from the point of view of direct experience, had nothing whatever to do with one another. The latter gave a mechanical interpretation of the thermodynamic ideas and laws as well as the discovery of the limit of applicability of the notions and laws to the classical theory of heat. This kinetic theory which surpassed, by far, the phenomenological physics as regards the logical unity of its foundations, produced moreover definite values for the true magnitudes of atoms and molecules which resulted from several independent methods and were thus placed beyond the realm of reasonable doubt. These decisive progresses were paid for by the coordination of atomistic entities to the material points, the constructively speculative character of which entities being obvious. Nobody could hope ever to "perceive directly" an atom. Laws concerning variables connected more directly with experimental facts (for example: temperature, pressure, speed) were deduced from the fundamental ideas by means of complicated calculations. In this manner physics (at least part of it), originally more phenomenologically constructed, was reduced, by being founded upon Newton's mechanics for atoms and molecules, to a basis further removed from direct experiment, but more uniform in character.

§ 3. The Field Concept

In explaining optical and electrical phenomena Newton's mechanics has been far less successful than it had been in the fields cited above. It is true that Newton tried to reduce light to the motion of material points in his corpuscular theory of light. Later on, however, as the phenomena of polarization, diffraction and interference of light forced upon his theory more and more unnatural modifications, Huyghens' undulatory theory of light, prevailed. Probably this theory owes its origin essentially to the phenomena of crystallographic optics and to the theory of sound, which was then already elaborated to a certain degree. It must be admitted that Huyghens' theory also was based in the first instance upon classical mechanics; but, the all-penetrating ether had to be assumed as the carrier of the waves and the structure of the ether, formed from material points, could not be explained by any known phenomenon. One could never get a clear picture of the interior forces governing the ether, nor of the forces acting between the ether and the "ponderable" matter. The foundations of this theory remained, therefore, eternally in the dark. The true basis was a partial differential equation, the reduction of which to mechanical elements remained always problematic.

For the theoretical conception of electric and magnetic phenomena one introduced, again, masses of a special kind, and between these masses one assumed the existence of forces acting at a distance, similar to Newton's gravitational forces. This special kind of matter, however, appeared to be lacking in the fundamental property of inertia; and, the forces acting between these masses and the ponderable matter remained obscure. To these difficulties there had to be added the polar character of these kinds of matter which did not fit into the scheme of classical mechanics. The basis of the theory became still more unsatisfactory when electrodynamic phenomena became known, notwithstanding the fact that these phenomena brought the physicist to the explanation of magnetic phenomena through electrodynamic phenomena and, in this way, made the assumption of magnetic masses superfluous. This progress had, indeed, to be paid for by increasing the complexity of the forces of interaction which had to be assumed as existing between electrical masses in motion.

The escape from this unsatisfactory situation by the electric field theory of Faraday and Maxwell represents probably the most profound transformation which has been experienced by the foundations of physics since Newton's time. Again, it has been a step in the direction of constructive speculation which has increased the distance between the foundation of the theory and what can be experienced by means of our five senses. The existence of the field manifests itself, indeed, only when electrically charged bodies are introduced into it. The differential equations of Maxwell connect the spacial and temporal differential coefficients of the electric and magnetic fields. The electric masses are nothing more than places of non-disappearing divergency of the electric field. Light waves appear as undulatory electromagnetic field processes in space.

To be sure, Maxwell still tried to interpret his field theory mechanically by means of mechanical ether models. But these attempts receded gradually to the background following the representation—purged of any unnecessary additions—by Heinrich Hertz, so that, in this theory the field finally took the fundamental position which had been occupied in Newton's mechanics by the material points. At first, however, this applies only for electromagnetic fields in empty space.

In its initial stage the theory was yet quite unsatisfactory for the interior of matter, because there, two electric vectors had to be introduced, which were connected by relations dependent on the nature of the medium, these relations being inaccessible to any theoretical analysis. An analogous situation arose in connection with the magnetic field, as well as in the relation between electric current density and the field.

Here H. A. Lorentz found an escape which showed, at the same time, the way to an electrodynamic theory of bodies in motion, a theory which was more or less free of arbitrary assumption. His theory was built on the following fundamental hypothesis:

Everywhere (including the interior of ponderable bodies) the seat of the field is the empty space. The participation of matter in electromagnetic phenomena has its origin only in the fact that the elementary particles of matter carry unalterable electric charges, and, on this account are subject on the one hand to the actions of ponderomotive forces and on the other hand possess the property of generating a field. The elementary particles obey Newton's law of motion for the material point.

This is the basis on which H. A. Lorentz obtained his synthesis of Newton's mechanics and Maxwell's field theory. The weakness of this theory lies in the fact that it tried to determine the phenomena by a combination of partial differential equations (Maxwell's field equations for empty space) and total differential equations (equations of motion of points), which procedure was obviously unnatural. The unsatisfactory part of the theory showed up externally by the necessity of assuming finite dimensions for the particles in order to prevent the electromagnetic field existing at their surfaces from becoming infinitely great. The theory failed moreover to give any explanation concerning the tremendous forces which hold the electric charges on the individual particles. H. A. Lorentz accepted these weaknesses of his theory, which were well known to him, in order to explain the phenomena correctly at least as regards their general lines.

Furthermore, there was one consideration which

reached beyond the frame of Lorentz's theory. In the environment of an electrically charged body there is a magnetic field which furnishes an (apparent) contribution to its inertia. Should it not be possible to explain the *total* inertia of the particles electromagnetically? It is clear that this problem could be worked out satisfactorily only if the particles could be interpreted as regular solutions of the electromagnetic partial differential equations. The Maxwell equations in their original form do not, however, allow such a description of particles, because their corresponding solutions contain a singularity. Theoretical physicists have tried for a long time, therefore, to reach the goal by a modification of Maxwell's equations. These attempts have, however, not been crowned with success. Thus it happened that the goal of erecting a pure electromagnetic field theory of matter remained unattained for the time being, although in principle no objection could be raised against the possibility of reaching such a goal. The thing which deterred one in any further attempt in this direction was the lack of any systematic method leading to the solution. What appears certain to me, however, is that, in the foundations of any consistent field theory, there shall not be, in addition to the concept of field, any concept concerning particles. The whole theory must be based solely on partial differential equations and their singularity-free solutions.

§ 4. The Theory of Relativity

There is no inductive method which could lead to the fundamental concepts of physics. Failure to understand this fact constituted the basic philosophical error of so many investigators of the nineteenth century. It was probably the reason why the molecular theory, and Maxwell's theory were able to establish themselves only at a relatively late date. Logical thinking is necessarily deductive; it is based upon hypothetical concepts and axioms. How can we hope to choose the latter in such a manner as to justify us in expecting success as a consequence?

The most satisfactory situation is evidently to be found in cases where the new fundamental hypotheses are suggested by the world of experience itself. The hypothesis of the non-existence of perpetual motion as a basis for thermodynamics affords such an example of a fundamental hypothesis suggested by experience; the same thing holds for the principle of inertia of Galileo. In the same category, moreover, we find the fundamental hypotheses of the theory of relativity, which theory has led to an unexpected expansion and broadening of the field theory, and to the superseding of the foundations of classical mechanics.

The successes of the Maxwell-Lorentz theory have given great confidence in the validity of the electromagnetic equations for empty space and hence, in particular, to the statement that light travels "in space" with a certain constant speed c. Is this law of the invariability of light velocity in relation to any desired inertial system valid? If it were not, then one specific inertial system or more accurately, one specific state of motion (of a body of reference), would be distinguished from all others. In opposition to this idea, however, stand all the mechanical and electromagnetic-optical facts of our experience.

For these reasons it was necessary to raise to the degree of a principle, the validity of the law of constancy of light velocity for all inertial systems. From this, it follows that the spacial coordinates X_1, X_2, X_3, and the time X_4, must be transformed according to the "Lorentz-transformation" which is characterized by invariance of the expression

$$ds^2 = dx_1^2 + dx_2^2 + dx_3^2 - dx_4^2$$

(if the unit of time is chosen in such a manner that the speed of light $c = 1$).

By this procedure time lost its absolute character, and was included with the "spacial" coordinates as of algebraically (nearly) similar character. The absolute character of time and particularly of simultaneity were destroyed, and the four dimensional description became introduced as the only adequate one.

In order to account, also, for the equivalence of all inertial systems with regard to all the phenomena of nature, it is necessary to postulate invariance of all systems of physical equations which express general laws, with regard to the Lorentzian transformation. The elaboration of this requirement forms the content of the special theory of relativity.

This theory is compatible with the equations of Maxwell; but, it is incompatible with the basis of classical mechanics. It is true that the equations of motion of the material point can be modified (and with them the expressions for momentum and kinetic energy of the material point) in such a manner as to satisfy the theory; but, the concept of the force of interaction, and with it the concept of potential energy of a system, lose their basis, because these concepts rest upon the idea of absolute instantaneousness. The field, as determined by differential equations, takes the place of the force.

Since the foregoing theory allows interaction only by fields, it requires a field theory of gravitation. Indeed, it is not difficult to formulate such a theory in which, as in Newton's theory, the gravitational fields can be reduced to a scalar which is the solution of a partial differential equation. However, the experimental facts expressed in Newton's theory of gravitation lead in another direction, that of the general theory of relativity.

Classical mechanics contains one point which is unsatisfactory in that, in the fundamentals, the same mass constant is met twice over in two different rôles, namely as "inertial mass" in the law of motion, and as "gravitational mass" in the law of gravitation. As a result of this, the acceleration of a body in a pure gravitational field is independent of its material; or, in a coordinate system of *uniform acceleration* (accelerated in relation to an "inertial system") the motions take place as they would in a homogeneous gravitational field (in relation to a "motionless" system of coordinates). If one assumes that the equivalence of these two cases is complete, then one attains an adaptation of our theoretical thinking to the fact that the gravitational and inertial masses are identical.

From this it follows that there is no longer any reason for favoring, as a fundamental principle, the "inertial systems"; and, we must admit as equivalent in their own right, also *non-linear* transformations of the coordinates (x_1, x_2, x_3, x_4). If we make such a transformation of a system of coordinates of the special theory of relativity, then the metric

$$ds^2 = dx_1{}^2 + dx_2{}^2 + dx_3{}^2 - dx_4{}^2$$

goes over to a general (Riemannian) metric of Bane

$$ds^2 = g_{\mu\nu}\, dx_\mu\, dx_\nu \quad \text{(Summed over } \mu \text{ and } \nu\text{)}$$

where the $g_{\mu\nu}$, symmetrical in μ and ν, are certain functions of $x_1 \ldots x_4$ which describe both the metric property, and the gravitational field in relation to the new system of coordinates.

The foregoing improvement in the interpretation of the mechanical basis must, however, be paid for in that —as becomes evident on closer scrutiny—the new coordinates could no longer be interpreted, as results of measurements by rigid bodies and clocks, as they could in the original system (an inertial system with vanishing gravitational field).

The passage to the general theory of relativity is realized by the assumption that such a representation of the field properties of space already mentioned, by functions $g_{\mu\nu}$ (that is to say by a Riemann metric), is also justified in the *general* case in which there is no system of coordinates in relation to which the metric takes the simple quasi-Euclidian form of the special theory of relativity.

Now the coordinates, by themselves, no longer express metric relations, but only the "neighborliness" of the things described, whose coordinates differ but little from one another. All transformations of the coordinates have to be admitted so long as these transformations are free from singularities. Only such equations as are covariant in relation to arbitrary transformations in this sense have meaning as expressions of general law of nature (postulate of general covariancy).

The first aim of the general theory of relativity was a preliminary statement which, by giving up the requirement of constituting a closed thing in itself, could be connected in as simple a manner as possible with the "facts directly observed." Newton's gravitational theory gave an example, by restricting itself to the pure mechanics of gravitation. This preliminary statement may be characterized as follows:

(1) The concept of the material point and of its mass is retained. A law of motion is given for it, this law of motion being the translation of the law of inertia into the language of the general theory of relativity. This law is a system of total differential equations, the system characteristic of the geodetic line.

(2) In place of Newton's law of interaction by gravitation, we shall find the system of the simplest generally covariant differential equations which can be set up for the $g_{\mu\nu}$-tensor. It is formed by equating to zero the once contracted Riemannian curvature tensor ($R_{\mu\nu} = 0$).

This formulation permits the treatment of the problem of the planets. More accurately speaking, it allows the treatment of the problem of motion of material points of practically negligible mass in the gravitational field produced by a material point which itself is supposed to have no motion (central symmetry). It does not take into account the reaction of the "moved" material points on the gravitational field, nor does it consider how the central mass produces this gravitational field.

Analogy with classical mechanics shows that the following is a way to complete the theory. One sets up as field equation

$$R_{ik} - \tfrac{1}{2}g_{ik}R = -T_{ik}$$

where R represents the scalar of Riemannian curvature, T_{ik} the energy tensor of the matter in a phenomenological representation. The left side of the equation is chosen in such a manner that its divergence disappears identically. The resulting disappearance of the divergence of the right side produces the "equations of motion" of matter, in the form of partial differential equations for the case where T_{ik} introduces, for the description of the matter, only *four* further functions independent of each other (for instance, density, pressure, and velocity components, where there is between the latter an identity, and between pressure and density an equation of condition).

By this formulation one reduces the whole mechanics of gravitation to the solution of a single system of covariant partial differential equations. The theory avoids all internal discrepancies which we have charged against the basis of classical mechanics. It is sufficient—as far as we know—for the representation of the observed facts of celestial mechanics. But, it is similar to a building, one wing of which is made of fine marble (left part of the equation), but the other wing of which is built of low grade wood (right side of equation). The phenomenological representation of matter is, in fact, only a crude substitute for a representation which would correspond to all known properties of matter.

There is no difficulty in connecting Maxwell's theory of the electromagnetic field with the theory of the gravitational field so long as one restricts himself to space, free of ponderable matter and free of electric density. All that is necessary is to put on the right hand side of the above equation for T_{ik}, the energy tensor of the electromagnetic field in empty space and to associate with the so modified system of equations the Maxwell field equation for empty space, written in general covariant form. Under these conditions there will exist, between all these equations, a sufficient number of the differential identities to guarantee their consistency. We may add that this necessary formal property of the total system of equations leaves arbitrary the choice of the sign of the member T_{ik}, a fact which was later shown to be important.

The desire to have, for the foundations of the theory, the greatest possible unity has resulted in several attempts to include the gravitational field and the electromagnetic field in one formal but homogeneous picture. Here we must mention particularly the five-dimensional theory of Kaluza and Klein. Having considered this possibility very carefully I feel that it is more desirable to accept the lack of internal uniformity of the original theory, because I do not consider that the totality of the hypothetical basis of the five-dimensional theory contains less of an arbitrary nature than does the original theory. The same statement may be made for the projective variety of the theory, which has been elaborated with great care, in particular, by v. Dantzig and by Pauli.

The foregoing considerations concern, exclusively, the theory of the field, free of matter. How are we to proceed from this point in order to obtain a complete theory of atomically constructed matter? In such a theory, singularities must certainly be excluded, since without such exclusion the differential equations do not completely determine the total field. Here, in the field theory of general relativity, we meet the same problem of a theoretical field-representation of matter as was met originally in connection with the pure Maxwell theory.

Here again the attempt to construct particles out of the field theory, leads apparently to singularities. Here also the endeavor has been made to overcome this defect by the introduction of new field variables and by elaborating and extending the system of field equations. Recently, however, I discovered, in collaboration with Dr. Rosen, that the above mentioned simplest combination of the field equations of gravitation and electricity produces centrally symmetrical solutions which can be represented as free of singularity (the well known centrally symmetrical solutions of Schwarzschild for the pure gravitational field, and those of Reissner for the electric field with consideration of its gravitational action). We shall refer to this shortly in the paragraph next but one. In this way it seems possible to get for matter and its interactions a pure field theory free of additional hypotheses, one moreover whose test by submission to facts of experience does not result in difficulties other than purely mathematical ones, which difficulties, however, are very serious.

§ 5. Quantum Theory and the Fundamentals of Physics

The theoretical physicists of our generation are expecting the erection of a new theoretical basis for physics which would make use of fundamental concepts greatly different from those of the field theory considered up to now. The reason is that it has been found necessary to use—for the mathematical representation of the so-called quantum phenomena—new sorts of methods of consideration.

While the failure of classical mechanics, as revealed by the theory of relativity, is connected with the finite speed of light (its avoidance of being ∞), it was discovered at the beginning of our century that there were other kinds of inconsistencies between deductions from mechanics and experimental facts, which inconsistencies are connected with the finite magnitude (the avoidance of being zero) of Planck's constant h. In particular, while molecular mechanics requires that both, heat content and (monochromatic) radiation density, of solid bodies should decrease *in proportion* to the decreasing absolute temperature, experience has shown that they decrease much more rapidly than the absolute temperature. For a theoretical explanation of this behavior it was necessary to assume that the energy of a mechanical system cannot assume any sort of value, but only certain discrete values whose mathematical expressions were always dependent upon Planck's constant h. Moreover, this conception was essential for the theory of the atom (Bohr's theory). For the transitions of these states into one another—with or without emission or absorption of radiation—no causal laws could be given, but only statistical ones; and, a similar conclusion holds for the radioactive decomposition of atoms, which decomposition was carefully investigated about the same time. For more than two decades physicists tried vainly to find a uniform interpretation of this "quantum character" of systems and phenomena. Such an attempt was successful about ten years ago, through the agency of two entirely different theoretical methods of attack. We owe one of these to Heisenberg and Dirac, and the other to de Broglie and Schrödinger. The mathematical equivalence of the two methods was soon recognized by Schrödinger. I shall try here to sketch the line of thought of de Broglie and Schrödinger, which lies closer to the physicist's method of thinking, and shall accompany the description with certain general considerations.

The question is first: How can one assign a discrete succession of energy value H_σ to a system specified in the sense of classical mechanics (the energy function is a given function of the coordinates q_r and the corresponding momenta p_r)? Planck's constant h relates the frequency H_σ/h to the energy values H_σ. It is therefore sufficient to give to the system a succession of discrete *frequency* values. This reminds us of the fact that in acoustics, a series of discrete frequency values is coordinated to a linear partial differential equation (if boundary values are given) namely the sinusoidal periodic solutions. In corresponding manner, Schrödinger set himself the task of coordinating a partial differential equation for a scalar function ψ to the given energy function $\mathcal{E}$ (q_r, p_r), where the q_r and the time t are independent variables. In this he succeeded (for a complex function ψ) in such a manner that the theoretical values of the energy H_σ, as required by the statistical theory, actually resulted in a satisfactory manner from the periodic solution of the equation.

To be sure, it did not happen to be possible to associate a definite movement, in the sense of mechanics of material points, with a definite solution ψ (q_r, t) of the Schrödinger equation. This means that the ψ function does not determine, at any rate *exactly*, the story of the q_r as functions of the time t. According to Born, however, an interpretation of the physical meaning of the ψ functions was shown to be possible in the following manner: $\psi\psi$ (the square of the absolute value of the complex function ψ) is the probability density at the point under consideration in the configuration-space of the q_r, at the time t. It is therefore possible to characterize the content of the Schrödinger equation in a manner, easy to be understood, but not quite accurate, as follows: it determines how the probability density of a statistical ensemble of systems varies in the configuration-space with the time. Briefly: the Schrödinger equation determines the alteration of the function ψ of the q_r with the time.

It must be mentioned that the result of this theory contains—as limiting values—the result of the particle mechanics if the wave-length encountered during the solution of the Schrödinger problem is everywhere so small that the potential energy varies by a practically infinitely small amount for a change of one wave-length in the configuration-space. Under these conditions the following can in fact be shown: We choose a region G_0 in the configuration-space which, although large (in every dimension) in relation to the wave length, is small in relation to the practical dimensions of the configuration-space. Under these conditions it is possible to choose a function of ψ for an initial time t_0 in such a manner that it vanishes outside of the region G_0, and behaves, according to the Schrödinger equation, in such a manner that it retains this property—approximately at least—also for a later time, but with the region G_0 having passed at that time t into another region G. In this manner one can, with a certain degree of approximation, speak of the motion of the region G as a whole, and one can approximate this motion by the motion of a point in the configuration-space. This motion then coincides with the motion which is required by the equations of classical mechanics.

Experiments on interference made with particle rays have given a brilliant proof that the wave character of phenomena of motion as assumed by the theory does, really, correspond to the facts. In addition to this, the theory succeeded, easily, in demonstrating the statistical laws of the transition of a system from one quantum condition to another under the action of external forces, which, from the standpoint of classical mechanics, appears as a miracle. The external forces were here represented by small additions of the potential energy as functions of the time. Now, while in classical mechanics, such additions can produce only correspondingly small alterations of the system, in the quantum mechanics they produce alterations of any magnitude however large, but with correspondingly small probability, a consequence in perfect harmony with experience. Even an understanding of the laws of radioactive decomposition, at least in their broad lines, was provided by the theory.

Probably never before has a theory been evolved which has given a key to the interpretation and calculation of such a heterogeneous group of phenomena of experience as has the quantum theory. In spite of this, however, I believe that the theory is apt to beguile us into error in our search for a uniform basis for physics, because, in my belief, it is an *incomplete* representation of real things, although it is the only one which can be built out of the fundamental concepts of force and material points (quantum corrections to classical mechanics). The incompleteness of the representation is the outcome of the statistical nature (incompleteness) of the laws. I will now justify this opinion.

I ask first: How far does the ψ function describe a real condition of a mechanical system? Let us assume the ψ_r to be the periodic solutions (put in the order of increasing energy values) of the Schrödinger equation. I shall leave open, for the time being, the question as to how far the individual ψ_r are *complete* descriptions of physical conditions. A system is first in the condition ψ_1 of lowest energy $\mathcal{E}_1$. Then during a finite time a small disturbing force acts upon the system. At a later instant one obtains then from the Schrödinger equation a ψ function of the form

$$\psi = \Sigma\, c_r \psi_r$$

where the c_r are (complex) constants. If the ψ_r are "normalized," then $|c_1|$ is nearly equal to 1, $|c_2|$ etc. is small compared with 1. One may now ask: Does ψ describe a real condition of the system? If the answer is yes, then we can hardly do otherwise than ascribe[4] to this condition a definite energy $\mathcal{E}$, and, in particular, such an energy as exceeds $\mathcal{E}_1$ by a small amount (in any case $\mathcal{E}_1 < \mathcal{E} < \mathcal{E}_2$). Such an assumption is, however, at variance with the experiments on electron impact such as have been made by J. Franck and G. Hertz, if, in addition to this, one accepts Millikan's demonstration of the discrete nature of electricity. As a matter of fact, these experiments lead to the conclusion that energy values of a state lying between the quantum values do not exist. From this it follows that our function ψ does not in any way describe a homogeneous condition of the body, but represents rather a statistical description in which the c_r represent probabilities of the individual energy values. It seems to be clear, therefore, that the Born statistical interpretation of the quantum theory is the only possible one. The ψ function does not in any way describe a condition which could be that of a single system; it relates rather to many systems, to "an ensemble of systems" in the sense of statistical mechanics. If, except for certain special cases, the ψ function furnishes only *statistical* data concerning measurable magnitudes, the reason lies not only in the fact that the *operation of measuring* introduces unknown elements, which can be grasped only statistically, but because of the very fact that the ψ function does not, in any sense, describe the condition of *one* single system. The Schrödinger equation determines the time variations which are experienced by the ensemble of systems which may exist with or without external action on the single system.

Such an interpretation eliminates also the paradox recently demonstrated by myself and two collaborators, and which relates to the following problem.

[4] Because, according to a well established consequence of the relativity theory, the energy of a complete system (at rest) is equal to its inertia (as a whole). This, however, must have a well defined value.

Consider a mechanical system constituted of two partial systems *A* and *B* which have interaction with each other only during limited time. Let the ψ function before their interaction be given. Then the Schrödinger equation will furnish the ψ function after the interaction has taken place. Let us now determine the physical condition of the partial system *A* as completely as possible by measurements. Then the quantum mechanics allows us to determine the ψ function of the partial system *B* from the measurements made, and from the ψ function of the total system. This determination, however, gives a result which depends upon *which* of the determining magnitudes specifying the condition of *A* has been measured (for instance coordinates *or* momenta). Since there can be only *one* physical condition of *B* after the interaction and which can reasonably not be considered as dependent on the particular measurement we perform on the system *A* separated from *B* it may be concluded that the ψ function is not unambiguously coordinated with the physical condition. This coordination of several ψ functions with the same physical condition of system *B* shows again that the ψ function cannot be interpreted as a (complete) description of a physical condition of a unit system. Here also the coordination of the ψ function to an ensemble of systems eliminates every difficulty.[5]

The fact that quantum mechanics affords, in such a simple manner, statements concerning (apparently) discontinuous transitions from one total condition to another without actually giving a representation of the specific process, this fact is connected with another, namely the fact that the theory, in reality, does not operate with the single system, but with a totality of systems. The coefficients c_r of our first example are really altered very little under the action of the external force. With this interpretation of quantum mechanics one can

[5] The operation of measuring *A*, for example, thus involves a transition to a narrower ensemble of systems. The latter (hence also its ψ function) depends upon the point of view according to which this narrowing of the ensemble of systems is made.

understand why this theory can easily account for the fact that weak disturbing forces are able to produce alterations of any magnitude in the physical condition of a system. Such disturbing forces produce, indeed, only correspondingly small alterations of the *statistical density* in the ensemble of systems, and hence only infinitely weak alterations of the ψ functions, the mathematical description of which offers far less difficulty than would be involved in the mathematical representation of finite alterations experienced by part of the single systems. What happens to the single system remains, it is true, entirely unclarified by this mode of consideration; this enigmatic happening is entirely eliminated from the representation by the statistical manner of consideration.

But now I ask: Is there really any physicist who believes that we shall never get any inside view of these important alterations in the single systems, in their structure and their causal connections, and this regardless of the fact that these single happenings have been brought so close to us, thanks to the marvelous inventions of the Wilson chamber and the Geiger counter? To believe this is logically possible without contradiction; but, it is so very contrary to my scientific instinct that I cannot forego the search for a more complete conception.

To these considerations we should add those of another kind which also voice their plea against the idea that the methods introduced by quantum mechanics are likely to give a useful basis for the whole of physics. In the Schrödinger equation, absolute time, and also the potential energy, play a decisive role, while these two concepts have been recognized by the theory of relativity as inadmissible in principle. If one wishes to escape from this difficulty he must found the theory upon field and field laws instead of upon forces of interaction. This leads us to transpose the statistical methods of quantum mechanics to fields, that is to systems of infinitely many degrees of freedom. Although the attempts so far made are restricted to linear equations, which, as we know from the results of the general theory of relativity, are insufficient, the complications met up to now by the very ingenious attempts are already terrifying. They certainly will rise sky high if one wishes to obey the requirements of the general theory of relativity, the justification of which in principle nobody doubts.

To be sure, it has been pointed out that the introduction of a space-time continuum may be considered as contrary to nature in view of the molecular structure of everything which happens on a small scale. It is maintained that perhaps the success of the Heisenberg method points to a purely algebraical method of description of nature, that is to the elimination of continuous functions from physics. Then, however, we must also give up, by principle, the space-time continuum.. It is not unimaginable that human ingenuity will some day find methods which will make it possible to proceed along such a path. At the present time, however, such a program looks like an attempt to breathe in empty space.

There is no doubt that quantum mechanics has seized hold of a beautiful element of truth, and that it will be a test stone for any future theoretical basis, in that it must be deducible as a limiting case from that basis, just as electrostatics is deducible from the Maxwell equations of the electromagnetic field or as thermodynamics is deducible from classical mechanics. However, I do not believe that quantum mechanics will be the *starting point* in the search for this basis, just as, vice versa, one could not go from thermodynamics (resp. statistical mechanics) to the foundations of mechanics.

In view of this situation, it seems to be entirely justifiable seriously to consider the question as to whether the basis of field physics cannot by *any* means be put into harmony with the facts of the quantum theory. Is this not the only basis which, consistently with today's possibility of mathematical expression, can be adapted to the requirements of the general theory of relativity? The belief, prevailing among the physicists of today, that such an attempt would be hopeless, may have its root in the unjustifiable idea that such a theory should lead, as a first approximation, to the equations of classical mechanics for the motion of corpuscles, or at least to total differential equations. As a matter of fact up to now we have never succeeded in representing corpuscles theoretically by fields free of singularities, and we can, a priori, say nothing about the behavior of such entities. *One thing,* however, is certain: if a field theory results in a representation of corpuscles free of singularities, then the behavior of these corpuscles with time is determined solely by the differential equations of the field.

§ 6. Relativity Theory and Corpuscles

I shall now show that, according to the general theory of relativity, there exist singularity-free solutions of field equations which can be interpreted as representing corpuscles. I restrict myself here to neutral particles because, in another recent publication in collaboration with Dr. Rosen, I have treated this question in a detailed manner, and because the essentials of the problem can be completely shown by this case.

The gravitational field is entirely described by the tensor $g_{\mu\nu}$. In the three-index symbols $\Gamma_{\mu\nu\sigma}$, there appear also the contravariants $g_{\mu\nu}$ which are defined as the minors of the $g_{\mu\nu}$ divided by the determinant $g\ (=|g_{\alpha\beta}|)$. In order that the R_{ik} shall be defined and finite, it is not sufficient that there shall be, for the environment of every part of the continuum, a system of coordinates in which the $g_{\mu\nu}$ and their first differential quotients are continuous and differentiable, but it is also necessary that the determinant g shall nowhere vanish. This last restriction is, however, eliminated if one replaces the differential equations $R_{ik} = 0$ by $g^2 R_{ik} = 0$, the left hand sides of which are *whole* rational functions of the g_{ik} and of their derivatives.

These equations have the centrally symmetrical solutions indicated by Schwarzschild

$$ds^2 = -\frac{1}{1 - 2m/r}dr^2 - r^2(d\theta^2 + \sin^2\theta d\varphi^2) + \left(1 - \frac{2m}{r}\right) dt^2$$

This solution has a singularity at $r = 2m$, since the coefficient of dr^2 (i.e. g_{11}), becomes infinite on this hypersurface. If, however, we replace the variable r by ρ defined by the equation

$$\rho^2 = r - 2m$$

we obtain

$$ds^2 = -4(2m + \rho^2)d\rho^2 - (2m + \rho^2)^2(d\theta^2 + \sin^2\theta d\varphi^2) + \frac{\rho^2}{2m + \rho^2} dt^2$$

This solution behaves regularly for all values of ρ. The vanishing of the coefficient of dt^2 i.e. (g_{44}) for $\rho = 0$ results, it is true, in the consequence that the determinant g vanishes for this value; but, with the methods of writing the field equations actually adopted, this does not constitute a singularity.

If ρ extends from $-\infty$ to $+\infty$, then r runs from $+\infty$ to $r = 2m$ and then back to $+\infty$, while for such values of r as correspond to $r < 2m$ there are no corresponding real values of ρ. Hence the Schwarzschild solution becomes a regular solution by representation of the physical space as consisting of two identical "shells" neighboring upon the hypersurface $\rho = 0$, that is $r = 2m$, while for this hypersurface the determinant g vanishes. Let us call such a connection between the two (identical) shells a "bridge." Hence the existence of such a bridge between the two shells in the finite realm corresponds to the existence of a material neutral particle which is described in a manner free from singularities.

The solution of the problem of the motion of neutral particles evidently amounts to the discovery of such solutions of the gravitational equations (written free of denominators), as contain several bridges.

The conception sketched above corresponds, a priori, to the atomistic structure of matter insofar as the "bridge" is by its nature a discrete element. Moreover, we see that the mass constant m of the neutral particles must necessarily be positive, since no solution free of singularities can correspond to the Schwarzschild solution for a negative value of m. Only the examination of the several-bridge-problem, can show whether or not this theoretical method furnishes an explanation of the empirically demonstrated equality of the masses of the particles found in nature, and whether it takes into account the facts which the quantum mechanics has so wonderfully comprehended.

In an analogous manner, it is possible to demonstrate that the combined equations of gravitation and electricity (with appropriate choice of the sign of the electrical member in the gravitational equations) produce a singularity-free bridge-representation of the electric corpuscle. The simplest solution of this kind is that for an electrical particle without gravitational mass.

So long as the important mathematical difficulties concerned with the solution of the several-bridge-problem, are not overcome, nothing can be said concerning the usefulness of the theory from the physicist's point of view. However, it constitutes, as a matter of fact, the first attempt towards the consistent elaboration of a field theory which presents a possibility of explaining the properties of matter. In favor of this attempt one should also add that it is based on the simplest possible relativistic field equations known today.

Summary

Physics constitutes a logical system of thought which is in a state of evolution, and whose basis cannot be obtained through distillation by any inductive method from the experiences lived through, but which can only be attained by free invention. The justification (truth content) of the system rests in the proof of usefulness of the resulting theorems on the basis of sense experiences, where the relations of the latter to the former can only be comprehended intuitively. Evolution is going on in the direction of increasing simplicity of the logical basis. In order further to approach this goal, we must make up our mind to accept the fact that the logical basis departs more and more from the facts of experience, and that the path of our thought from the fundamental basis to these resulting theorems, which correlate with sense experiences, becomes continually harder and longer.

Our aim has been to sketch, as briefly as possible, the development of the fundamental concepts in their dependence upon the facts of experience and upon the strife towards the goal of internal perfection of the system. Today's state of affairs had to be illuminated by these considerations, as they appear to me. (It is unavoidable that historic schematic representation is of a personal color.)

I try to demonstrate how the concepts of bodily objects, space, subjective and objective time, are connected with one another and with the nature of the experience. In classical mechanics the concepts of space and time become independent. The concept of the bodily object is replaced in the foundations by the concept of the material point, by which means mechanics becomes fundamentally atomistic. Light and electricity produce insurmountable difficulties when one attempts to make mechanics the basis of all physics. We are thus led to the field theory of electricity, and, later on to the attempt to base physics entirely upon the concept of the field (after an attempted compromise with classical mechanics). This attempt leads to the theory of relativity (evolution of the notion of space and time into that of the continuum with metric structure).

I try to demonstrate, furthermore, why in my opinion the quantum theory does not seem likely to be able to produce a usable foundation for physics: one becomes involved in contradictions if one tries to consider the theoretical quantum description as a *complete* description of the individual physical system or happening.

On the other hand, up to the present time, the field theory is unable to give an explanation of the molecular structure of matter and of quantum phenomena. It is shown, however, that the conviction to the effect that the field theory is unable to give, by its methods, a solution of these problems rests upon prejudice.

4. *THE FUNDAMENTALS OF THEORETICAL PHYSICS*

SCIENCE IS THE ATTEMPT to make the chaotic diversity of our sense-experience correspond to a logically uniform system of thought. In this system single experiences must be correlated with the theoretic structure in such a way that the resulting coordination is unique and convincing.

The sense-experiences are the given subject-matter. But the theory that shall interpret them is man-made. It is the result of an extremely laborious process of adaptation: hypothetical, never completely final, always subject to question and doubt.

The scientific way of forming concepts differs from that which we use in our daily life, not basically, but merely in the more precise definition of concepts and conclusions; more painstaking and systematic choice of experimental material; and greater logical economy. By this last we mean the effort to reduce all concepts and correlations to as few as possible logically independent basic concepts and axioms.

What we call physics comprises that group of natural sciences which base their concepts on measurements; and whose concepts and propositions lend themselves to mathematical formulation. Its realm is accordingly defined as that part of the sum total of our knowledge which is capable of being expressed in mathematical terms. With the progress of science, the realm of physics has so expanded that it seems to be limited only by the limitations of the method itself.

The larger part of physical research is devoted to the development of the various branches of physics, in each of which the object is the theoretical understanding of more or less restricted fields of experience, and in each of which the laws and concepts remain as closely as possible related to experience. It is this department of science, with its ever-growing specialization, which has revolutionized practical life in the last centuries, and given birth to the possibility that man may at last be freed from the burden of physical toil.

On the other hand, from the very beginning there has always been present the attempt to find a unifying theoretical basis for all these single sciences, consisting of a minimum of concepts and fundamental relationships, from which all the concepts and relationships of the single disciplines might be derived by logical process. This is what we mean by the search for a foundation of the whole of physics. The confident belief that this ultimate goal may be reached is the chief source of the passionate devotion which has always animated the researcher. It is in this sense that the following observations are devoted to the foundations of physics.

From what has been said it is clear that the word foundations in this connection does not mean something analogous in all respects to the foundations of a building. Logically considered, of course, the various single laws of physics rest upon this foundation. But whereas a building may be seriously damaged by a heavy storm or spring flood, yet its foundations remain intact, in science the logical foundation is always in greater peril from new experiences or new knowledge than are the branch disciplines with their closer experimental contacts. In the connection of the foundation with all the single parts lies its great significance, but likewise its greatest danger in face of any new factor. When we realize this, we are led to wonder why the so-called revolutionary epochs of the science of physics have not more often and more completely changed its foundation than has actually been the case.

The first attempt to lay a uniform theoretical foundation was the work of Newton. In his system everything is reduced to the following concepts: (1) Mass points with invariable mass; (2) action at a distance between any pair of mass points; (3) law of motion for the mass point. There was not, strictly speaking, any all-embracing foundation, because an explicit law was formulated only for the actions-at-a-distance of gravitation; while for other actions-at-a-distance nothing was established *a priori* except the law of equality of *actio* and *reactio*. Moreover, Newton himself fully realized that time and space were essential elements, as physically effective factors, of his system, if only by implication.

This Newtonian basis proved eminently fruitful and was regarded as final up to the end of the nineteenth century. It not only gave results for the movements of the heavenly bodies, down to the most minute details, but also furnished a theory of the mechanics of discrete and continuous masses, a simple explanation of the principle of the conservation of energy and a complete and brilliant theory of heat. The explanation of the facts of electrodynamics within the Newtonian system was more forced; least convincing of all, from the very beginning, was the theory of light.

It is not surprising that Newton would not listen to a wave theory of light; for such a theory was most unsuited to his theoretical foundation. The assumption that space was filled with a medium consisting of material points that propagated light waves without exhibiting any other mechanical properties must have seemed to him quite artificial. The strongest empirical arguments for the wave nature of light, fixed speeds of propagation, interference, diffraction, polarization, were either unknown or else not known in any well-ordered synthesis. He was justified in sticking to his corpuscular theory of light.

During the nineteenth century the dispute was settled in favor of the wave theory. Yet no serious doubt of the mechanical foundation of physics arose, in the first place because nobody knew where to find a foundation of another sort. Only slowly, under the irresistible pressure of facts, there developed a new foundation of physics, field-physics.

From Newton's time on, the theory of action-at-a-distance was constantly found artificial. Efforts were not lacking to explain gravitation by a kinetic theory, that is, on the basis of collision forces of hypothetical mass particles. But the attempts were superficial and bore no fruit. The strange part played by space (or the inertial system) within the mechanical foundation was also clearly recognized, and criticized with especial clarity by Ernst Mach.

The great change was brought about by Faraday, Maxwell and Hertz—as a matter of fact half-unconsciously and against their will. All three of them, throughout their lives, considered themselves adherents of the mechanical theory. Hertz had found the simplest form of the equations of the electromagnetic field, and declared that any theory leading to these equations was Maxwellian theory. Yet toward the end of his short life he wrote a paper in which he presented as the foundation of physics a mechanical theory freed from the force-concept.

For us, who took in Faraday's ideas so to speak with our mother's milk, it is hard to appreciate their greatness and audacity. Faraday must have grasped with unerring instinct the artificial nature of all attempts to refer electromagnetic phenomena to actions-at-a-distance between electric particles reacting on each other. How was each single iron filing among a lot scattered on a piece of paper to know of the single electric particles running round in a nearby conductor? All these electric particles together seemed to create in the surrounding space a condition which in turn produced a certain order in the filings. These spatial states, to-day called fields, if their geometrical structure and interdependent action were once rightly grasped, would, he was convinced, furnish the clue to the mysterious electromagnetic interactions. He conceived these fields as states of mechanical stress in a space-filling medium, similar to the states of stress in an elastically distended body. For at that time this was the only way one could conceive of states that were apparently continuously distributed in space. The peculiar type of mechanical interpretation of these fields remained in the background—a sort of placation of the scientific conscience in view of the mechanical tradition of Faraday's time. With the help of these new field concepts Faraday succeeded in forming a qualitative concept of the whole complex of electromagnetic effects discovered by him and his predecessors. The precise formulation of the time-space laws of those fields was the work of Maxwell. Imagine his feelings when the differential equations he had formulated proved to him that electromagnetic fields spread in the form of polarized waves and with the speed of light! To few men in the world has such an experience been vouchsafed. At that thrilling moment he surely never guessed that the riddling nature of light, apparently so completely solved, would continue to baffle succeeding generations. Meantime, it took physicists some decades to grasp the full significance of Maxwell's discovery, so bold was the leap that his genius forced upon the conceptions of his fellow-workers. Only after Hertz had demonstrated experimentally the existence of Maxwell's electromagnetic waves, did resistance to the new theory break down.

But if the electromagnetic field could exist as a wave independent of the material source, then the electrostatic interaction could no longer be explained as action-at-a-distance. And what was true for electrical action could not be denied for gravitation. Everywhere Newton's actions-at-a-distance gave way to fields spreading with finite velocity.

Of Newton's foundation there now remained only the material mass points subject to the law of motion. But J. J. Thomson pointed out that an electrically charged body in motion must, according to Maxwell's theory, possess a magnetic field whose energy acted precisely as does an increase of kinetic energy to the body. If, then, a part of kinetic energy consists of field energy, might that not then be true of the whole of the kinetic energy? Perhaps the basic property of matter, its inertia, could be explained within the field theory? The question led to the problem of an interpretation of matter in terms of field theory, the solution of which would furnish an explanation of the atomic structure of matter. It was soon realized that Maxwell's theory could not accomplish such a program. Since then many scientists have zealously sought to complete the field theory by some generalization that should comprise a theory of matter; but so far such efforts have not been crowned with success. In order to construct a theory, it is not enough to have a clear conception of the goal. One must also have a formal point of view which will sufficiently restrict the unlimited variety of possibilities. So far this has not been found; accordingly the field theory has not succeeded in furnishing a foundation for the whole of physics.

For several decades most physicists clung to the conviction that a mechanical substructure would be found for Maxwell's theory. But the unsatisfactory results of their efforts led to gradual acceptance of the new field concepts as irreducible fundamentals—in other words, physicists resigned themselves to giving up the idea of a mechanical foundation.

Thus physicists held to a field-theory program. But it could not be called a foundation, since nobody could tell whether a consistent field theory could ever explain on the one hand gravitation, on the other hand the elementary components of matter. In this state of affairs it was necessary to think of material particles as mass points subject to Newton's laws of motion. This was the procedure of Lorentz in creating his electron theory and the theory of the electromagnetic phenomena of moving bodies.

Such was the point at which fundamental conceptions had arrived at the turn of the century. Immense progress was made in the theoretical penetration and understanding of whole groups of new phenomena; but the establishment of a unified foundation for physics seemed remote indeed. And this state of things has even been aggravated by subsequent developments. The development during the present century is characterized by two theoretical systems essentially independent of each other: the theory of relativity and the quantum theory. The two systems do not directly contradict each other; but they seem little adapted to fusion into one unified theory. We must briefly discuss the basic idea of these two systems.

The theory of relativity arose out of efforts to improve, with reference to logical economy, the foundation of physics as it existed at the turn of the century. The so-called special or restricted relativity theory is based on the fact that Maxwell's equations (and thus the law of propagation of light in empty space) are converted into equations of the same form, when they undergo Lorentz transformation. This formal property of the Maxwell equations is supplemented by our fairly secure empirical knowledge that the laws of physics are the same with respect to all inertial systems. This leads to the result that the Lorentz transformation—applied to space and time coordinates—must govern the transition from one inertial system to any other. The content of the restricted relativity theory can accordingly be summarized in one sentence: all natural laws must be so conditioned that they are covariant with respect to Lorentz transformations. From this it follows that the simultaneity of two distant events is not an invariant concept and that the dimensions of rigid bodies and the speed of clocks depend upon their state of motion. A further consequence was a modification of Newton's law of motion in cases where the speed of a given body was not small compared with the speed of light. There followed also the principle of the equivalence of mass and energy, with the laws of conservation of mass and energy becoming one and the same. Once it was shown that simultaneity was relative and depended on the frame of reference, every possibility of retaining actions-at-a-distance within the foundation of physics disappeared, since that concept presupposed the absolute character of simultaneity (it must be possible to state the location of the two interacting mass points "at the same time").

The general theory of relativity owes its origin to the attempt to explain a fact known since Galileo's and Newton's time but hitherto eluding all theoretical interpretation: the inertia and the weight of a body, in themselves two entirely distinct things, are measured by one and the same constant, the mass. From this correspondence follows that it is impossible to discover by experiment whether a given system of coordinates is accelerated, or whether its motion is straight and uniform and the observed effects are due to a gravitational field (this is the equivalence principle of the general relativity theory). It shatters the concepts of the inertial system, as soon as gravitation enters in. It may be remarked here that the inertial system is a weak point of the Galilean-Newtonian mechanics. For there is presupposed a mysterious property of physical space, conditioning the kind of coordination-systems for which the law of inertia and the Newtonian law of motion hold good.

These difficulties can be avoided by the following postulate: natural laws are to be formulated in such a way that their form is identical for coordinate systems of any kind of states of motion. To accomplish this is the task of the general theory of relativity. On the other hand, we deduce from the restricted theory the existence of a Riemannian metric within the time-space continuum, which, according to the equivalence principle, describes both the gravitational field and the metric properties of space. Assuming that the field equations of gravitation are of the second differential order, the field law is clearly determined.

Aside from this result, the theory frees field physics from the disability it suffered from, in common with the Newtonian mechanics, of ascribing to space those independent physical properties which heretofore had been concealed by the use of an inertial system. But it can not be claimed that those parts of the general relativity theory which can to-day be regarded as final have furnished physics with a complete and satisfactory foundation. In the first place, the total field appears in it to be composed of two logically unconnected parts, the gravitational and the electromagnetic. And in the second place, this theory, like the earlier field theories, has not up till now supplied an explanation of the atomistic structure of matter. This failure has probably some connection with the fact that so far it has contributed nothing to the understanding of quantum phenomena. To take in these phenomena, physicists have been driven to the adoption of entirely new methods, the basic characteristics of which we shall now discuss.

In the year nineteen hundred, in the course of a purely theoretic investigation, Max Planck made a very remarkable discovery: the law of radiation of bodies as a function of temperature could not be derived solely from the laws of Maxwellian electrodynamics. To arrive at results consistent with the relevant experiments, radiation of a given frequency had to be treated as though it consisted of energy atoms of the individual energy h.v., where h is Planck's universal constant. During the years following it was shown that light was everywhere produced and absorbed in such energy quanta. In particular Niels Bohr was able largely to understand the structure of the atom, on the assumption that atoms can have only discrete energy values, and that the discontinuous transitions between them are connected with the emission or absorption of such an energy quantum. This threw some light on the fact that in their gaseous state elements and their compounds radiate and absorb only light of certain sharply defined frequencies. All this was quite inexplicable within the frame of the hitherto existing theories. It was clear that at least in the field of atomistic phenomena the character of everything that happens is determined by discrete states and by apparently discontinuous transitions between them, Planck's constant h playing a decisive role.

The next step was taken by De Broglie. He asked himself how the discrete states could be understood by the aid of the current concepts, and hit on a parallel with stationary waves, as for instance in the case of the proper frequencies of organ pipes and strings in acoustics. True, wave actions of the kind here required were unknown; but they could be constructed, and their mathematical laws formulated, employing Planck's constant h. De Broglie conceived an electron revolving about the atomic nucleus as being connected with such a hypothetical wave train, and made intelligible to some extent the discrete character of Bohr's "permitted" paths by the stationary character of the corresponding waves.

Now in mechanics the motion of material points is determined by the forces or fields of force acting upon them. Hence it was to be expected that those fields of force would also influence De Broglie's wave fields in an analogous way. Erwin Schrödinger showed how this influence was to be taken into account, re-interpreting by an ingenious method certain formulations of classical mechanics. He even succeeded in expanding the wave mechanical theory to a point where without the introduction of any additional hypotheses, it became applicable to any mechanical system consisting of an arbitrary number of mass points, that is to say possessing an arbitrary number of degrees of freedom. This was possible because a mechanical system consisting of n mass points is mathematically equivalent to a considerable degree, to one single mass point moving in a space of 3 n dimensions.

On the basis of this theory there was obtained a surprisingly good representation of an immense variety of facts which otherwise appeared entirely incomprehensible. But on one point, curiously enough, there was failure: it proved impossible to associate with these Schrödinger waves definite motions of the mass points—and that, after all, had been the original purpose of the whole construction.

The difficulty appeared insurmountable, until it was overcome by Born in a way as simple as it was unexpected. The De Broglie-Schrödinger wave fields were not to be interpreted as a mathematical description of how an event actually takes place in time and space, though, of course, they have reference to such an event. Rather they are a mathematical description of what we can actually know about the system. They serve only to make statistical statements and predictions of the results of all measurements which we can carry out upon the system.

Let me illustrate these general features of quantum mechanics by means of a simple example: we shall consider a mass point kept inside a restricted region G by forces of finite strength. If the kinetic energy of the mass point is below a certain limit, then the mass point, according to classical mechanics, can never leave the region G. But according to quantum mechanics, the mass point, after a period not immediately predictable, is able to leave the region G, in an unpredictable direction, and escape into surrounding space. This case, according to Gamow, is a simplified model of radioactive disintegration.

The quantum theoretical treatment of this case is as follows: at the time t_0 we have a Schrödinger wave system entirely inside G. But from the time t_0 onwards, the waves leave the interior of G in all directions, in such a way that the amplitude of the outgoing wave is small compared to the initial amplitude of the wave system inside G. The further these outside waves spread, the more the amplitude of the waves inside G diminishes, and correspondingly the intensity of the later waves issuing from G. Only after infinite time has passed is the wave supply inside G exhausted, while the outside wave has spread over an ever-increasing space.

But what has this wave process to do with the first object of our interest, the particle originally enclosed in G? To answer this question, we must imagine some arrangement which will permit us to carry out measurements on the particle. For instance, let us imagine somewhere in the surrounding space a screen so made that the particle sticks to it on coming into contact with it. Then from the intensity of the waves hitting the screen at some point, we draw conclusions as to the probability of the particle hitting the screen there at that time. As soon as the particle has hit any particular point of the screen, the whole wave field loses all its physical meaning; its only purpose was to make probability predictions as to the place and time of the particle hitting the screen (or, for instance, its momentum at the time when it hits the screen).

All other cases are analogous. The aim of the theory is to determine the probability of the results of measurement upon a system at a given time. On the other hand, it makes no attempt to give a mathematical representation of what is actually present or goes on in space and time. On this point the quantum theory of to-day differs fundamentally from all previous theories of physics, mechanistic as well as field theories. Instead of a model description of actual space-time events, it gives the probability distributions for possible measurements as functions of time.

It must be admitted that the new theoretical conception owes its origin not to any flight of fancy but to the compelling force of the facts of experience. All attempts to represent the particle and wave features displayed in the phenomena of light and matter, by direct course to a space-time model, have so far ended in failure. And Heisenberg has convincingly shown, from an empirical point of view, any decision as to a rigorously deterministic structure of nature is definitely ruled out, because of the atomistic structure of our experimental apparatus. Thus it is probably out of the question that any future knowledge can compel physics again to relinquish our present statistical theoretical foundation in favor of a deterministic one which would deal directly with physical reality. Logically the problem seems to offer two possibilities, between which we are in principle given a choice. In the end the choice will be made according to which kind of description yields the formulation of the simplest foundation, logically speaking. At the present, we are quite without any deterministic theory directly describing the events themselves and in consonance with the facts.

For the time being, we have to admit that we do not possess any general theoretical basis for physics, which can be regarded as its logical foundation. The field theory, so far, has failed in the molecular sphere. It is agreed on all hands that the only principle which could serve as the basis of quantum theory would be one that constituted a translation of the field theory into the scheme of quantum statistics. Whether this will actually come about in a satisfactory manner, nobody can venture to say.

Some physicists, among them myself, can not believe that we must abandon, actually and forever, the idea of direct representation of physical reality in space and time; or that we must accept the view that events in nature are analogous to a game of chance. It is open to every man to choose the direction of his striving; and also every man may draw comfort from Lessing's fine saying, that the search for truth is more precious than its possession.

5. *THE COMMON LANGUAGE OF SCIENCE*

THE FIRST STEP towards language was to link acoustically or otherwise commutable signs to sense-impressions. Most likely all sociable animals have arrived at this primitive kind of communication—at least to a certain degree. A higher development is reached when further signs are introduced and understood which establish relations between those other signs designating sense-impression. At this stage it is already possible to report somewhat complex series of impressions; we can say that language has come to existence. If language is to lead at all to understanding, there must be rules concerning the relations between the signs on the one hand and on the other hand there must be a stable correspondence between signs and impressions. In their childhood individuals connected by the same language grasp these rules and relations mainly by intuition. When man becomes conscious of the rules concerning the relations between signs the so-called grammar of language is established.

In an early stage the words may correspond directly to impressions. At a later stage this direct connection is lost insofar as some words convey relations to perceptions only if used in connection with other words (for instance such words as: "is," "or," "thing"). Then word-groups rather than single words refer to perceptions. When language becomes thus partially independent from the background of impressions a greater inner coherence is gained.

Only at this further development where frequent use is made of so-called abstract concepts, language becomes an instrument of reasoning in the true sense of the word. But it is also this development which turns language into a dangerous source of error and deception. Everything depends on the degree to which words and word-combinations correspond to the world of impression.

What is it that brings about such an intimate connection between language and thinking? Is there no thinking without the use of language, namely in concepts and concept-combinations for which words need not necessarily come to mind? Has not everyone of us struggled for words although the connection between "things" was already clear?

We might be inclined to attribute to the act of thinking complete independence from language if the individual formed or were able to form his concepts without the verbal guidance of his environment. Yet most likely the mental shape of an individual, growing up under such conditions, would be very poor. Thus we may conclude that the mental development of the individual and his way of forming concepts depend to a high degree upon language. This makes us realize to what extent the same language means the same mentality. In this sense thinking and language are linked together.

What distinguishes the language of science from language as we ordinarily understand the word? How is it that scientific language is international? What science strives for is an utmost acuteness and clarity of concepts as regards their mutual relation and their correspondence to sensory data. As an illustration let us take the language of Euclidian geometry and Algebra. They manipulate with a small number of independently introduced concepts, respectively symbols, such as the integral number, the straight line, the point, as well as with signs which designate the fundamental operations, that is the connections between those fundamental concepts. This is the basis for the construction, respectively definition of all other statements and concepts. The connection between concepts and statements on the one hand and the sensory data on the other hand is established through acts of counting and measuring whose performance is sufficiently well determined.

The super-national character of scientific concepts and scientific language is due to the fact that they have been set up by the best brains of all countries and all times. In solitude and yet in cooperative effort as regards the final effect they created the spiritual tools for the technical revolutions which have transformed the life of mankind in the last centuries. Their system of concepts have served as a guide in the bewildering chaos of perceptions so that we learned to grasp general truths from particular observations.

What hopes and fears does the scientific method imply for mankind? I do not think that this is the right way to put the question. Whatever this tool in the hand of man will produce depends entirely on the nature of the goals alive in this mankind. Once these goals exist, the scientific method furnishes means to realize them. Yet it cannot furnish the very goals. The scientific method itself would not have led anywhere, it would not even have been born without a passionate striving for clear understanding.

Perfections of means and confusion of goals seem—in my opinion—to characterize our age. If we desire sincerely and passionately the safety, the welfare and the free development of the talents of all men, we shall not be in want of the means to approach such a state. Even if only a small part of mankind strives for such goals, their superiority will prove itself in the long run.

6. *THE LAWS OF SCIENCE AND THE LAWS OF ETHICS*

SCIENCE SEARCHES FOR RELATIONS which are thought to exist independently of the searching individual. This includes the case where man himself is the subject. Or the subject of scientific statements may be concepts created by ourselves, as in mathematics. Such concepts are not necessarily supposed to correspond to any objects in the outside world. However, all scientific statements and laws have one characteristic in common: they are "true or false" (adequate or inadequate). Roughly speaking, our reaction to them is "yes" or "no."

The scientific way of thinking has a further characteristic. The concepts which it uses to build up its coherent systems are not expressing emotions. For the scientist, there is only "being," but no wishing, no valuing, no good, no evil; no goal. As long as we remain within the realm of science proper, we can never meet with a sentence of the type: "Thou shalt not lie." There is something like a Puritan's restraint in the scientist who seeks truth: he keeps away from everything voluntaristic or emotional. Incidentally, this trait is the result of a slow development, peculiar to modern Western thought.

From this it might seem as if logical thinking were irrelevant for ethics. Scientific statements of facts and relations, indeed, cannot produce ethical directives. However, ethical directives can be made rational and coherent by logical thinking and empirical knowledge. If we can agree on some fundamental ethical propositions, then other ethical propositions can be derived from them, provided that the original premises are stated with sufficient precision. Such ethical premises play a similar role in ethics, to that played by axioms in mathematics.

This is why we do not feel at all that it is meaningless to ask such questions as: "Why should we not lie?" We feel that such questions are meaningful because in all discussions of this kind some ethical premises are tacitly taken for granted. We then feel satisfied when we succeed in tracing back the ethical directive in question to these basic premises. In the case of lying this might perhaps be done in some way such as this: Lying destroys confidence in the statements of other people. Without such confidence, social cooperation is made impossible or at least difficult. Such cooperation, however, is essential to make human life possible and tolerable. This means that the rule "Thou shalt not lie" has been traced back to the demands: "Human life shall be preserved" and "Pain and sorrow shall be lessened as much as possible."

But what is the origin of such ethical axioms? Are they arbitrary? Are they based on mere authority? Do they stem from experiences of men and are they conditioned indirectly by such experiences?

For pure logic all axioms are arbitrary, including the axioms of ethics. But they are by no means arbitrary from a psychological and genetic point of view. They are derived from our inborn tendencies to avoid pain and annihilation, and from the accumulated emotional reaction of individuals to the behavior of their neighbors.

It is the privilege of man's moral genius, impersonated by inspired individuals, to advance ethical axioms which are so comprehensive and so well founded that men will accept them as grounded in the vast mass of their individual emotional experiences. Ethical axioms are found and tested not very differently from the axioms of science. Truth is what stands the test of experience.

7. *AN ELEMENTARY DERIVATION OF THE EQUIVALENCE OF MASS AND ENERGY*

THE FOLLOWING DERIVATION of the law of equivalence, which has not been published before, has two advantages. Although it makes use of the principle of special relativity, it does not presume the formal machinery of the theory but uses only three previously known laws:

(1) The law of the conservation of momentum.
(2) The expression for the pressure of radiation; that is, the momentum of a complex of radiation moving in a fixed direction.
(3) The well known expression for the aberration of light (influence of the motion of the earth on the apparent location of the fixed stars—Bradley).

We now consider the following system. Let the body B rest freely in space with respect to the system K_0. Two

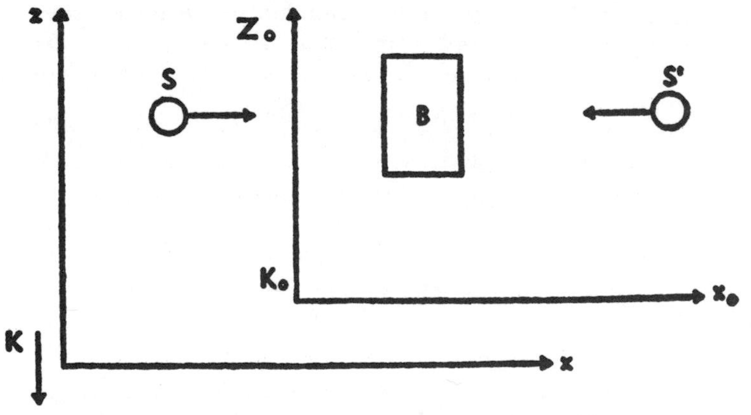

complexes of radiation S, S′ each of energy $\frac{E}{2}$ move in the positive and negative x_0 direction respectively and are eventually absorbed by B. With this absorption the energy of B increases by E. The body B stays at rest with respect to K_0 by reasons of symmetry.

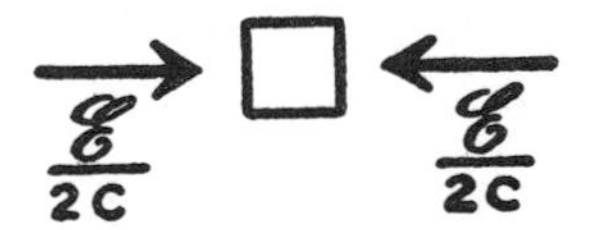

Now we consider this same process with respect to the system K, which moves with respect to K_0 with the constant velocity v in the negative Z_0 direction. With respect to K the description of the process is as follows:

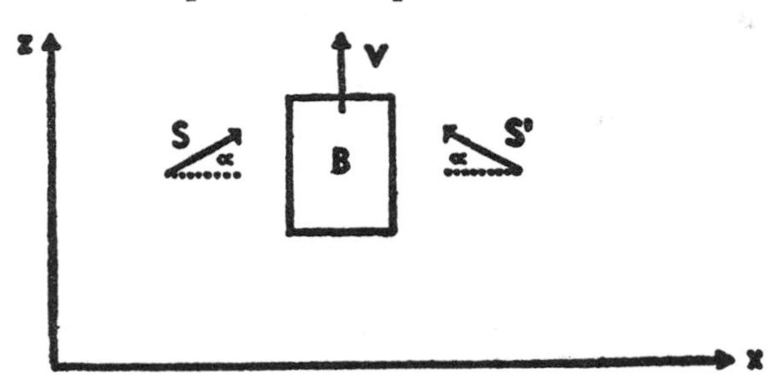

The body B moves in the positive Z direction with velocity v. The two complexes of radiation now have directions with respect to K which make an angle α with the x axis. The law of aberration states that in the first approximation $\alpha = \frac{c}{v}$, where c is the velocity of light. From the consideration with respect to K_0 we know that the velocity v of B remains unchanged by the absorption of S and S′.

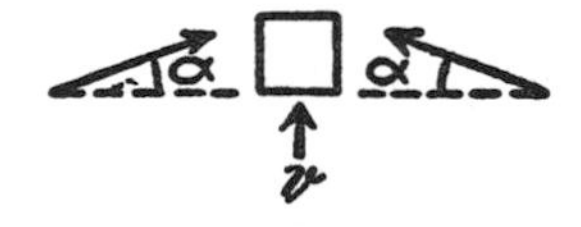

Now we apply the law of conservation of momentum with respect to the z direction to our system in the coordinate-frame K.

I. *Before the absorption* let M be the mass of B; Mv

is then the expression of the momentum of B (according to classical mechanics). Each of the complexes has the energy $\frac{E}{2}$ and hence, by a well known conclusion of Maxwell's theory, it has the momentum $\frac{E}{2c}$. Rigorously speaking this is the momentum of S with respect to K_0. However, when v is small with respect to c, the momentum with respect to K is the same except for a quantity of second order of magnitude $\left(\frac{v^2}{c^2} \text{ compared to } 1\right)$. The z-component of this momentum is $\frac{E}{2c}\sin\alpha$ or with sufficient accuracy (except for quantities of higher order of magnitude) $\frac{E}{2c}\alpha$ or $\frac{E}{2} \cdot \frac{v}{c^2}$. S and S′ together therefore have a momentum $E\frac{v}{c^2}$ in the z direction. The total momentum of the system before absorption is therefore

$$Mv + \frac{E}{c^2} \cdot v$$

II. *After the absorption* let M′ be the mass of B. We anticipate here the possibility that the mass increased with the absorption of the energy E (this is necessary so that the final result of our consideration be consistent). The momentum of the system after absorption is then

$$M'v$$

We now assume the law of the conservation of momentum and apply it with respect to the z direction. This gives the equation

$$Mv + \frac{E}{c^2}v = M'v$$

or

$$M' - M = \frac{E}{c^2}$$

This equation expresses the law of the equivalence of energy and mass. The energy increase E is connected with the mass increase $\frac{E}{c^2}$. Since energy according to the usual definition leaves an additive constant free, we may so choose the latter that

$$E = Mc^2$$

ACKNOWLEDGMENTS

1. From *The American People's Encyclopedia,* copyright by the Spencer Press, Inc., Chicago, 1949.
2. From *Science Illustrated;* New York, April, 1946.
3. From *The Journal of the Franklin Institute,* Vol. 221, No. 3; March, 1936.
4. From *Science;* Washington, D.C., May 24, 1940.
5. A broadcast recording for the Science Conference; London, September 28, 1941, and published in *Advancement of Science;* London, Vol. 2, no. 5.
6. From *Relativity—A Richer Truth* by Philipp Frank; published by the Beacon Press, Boston, 1950.
7. From *Technion Journal;* New York, 1946.